RIGHTSTART™ MATHEMATICS

by Joan A. Cotter, Ph.D.
with Tracy Mittleider, MSEd

LEVEL C WORKSHEETS

Second Edition

Activities for Learning, Inc.

A special thank you to Kathleen Cotter Clayton for all her work on the preparation of this manual.

Printed in the United States of America

www.RightStartMath.com

For more information: info@RightStartMath.com
Supplies may be ordered from: www.RightStartMath.com

Activities for Learning, Inc.
321 Hill Street
Hazelton, ND 58544-0468
United States of America
888-775-6284 or 701-782-2000
701-782-2007 fax

ISBN 978-1-931980-71-5
May 2024

Game Log

Math games are an important part of the RightStart™ Mathematics program. Instead of flash-cards or timed tests, games supply the necessary repetition for understanding and mastery in an enjoyable setting, providing a superior educational atmosphere. The lessons indicate which games to play. Additional games can be found in the *Math Card Games* book.

This Game Log provides a record of the games played and when they were played.

Date	Game Played	Players

Games help you understand, apply, and enjoy mathematics.

GAME LOG CONTINUED

Date	Game Played	Players

Games help you understand, apply, and enjoy mathematics.

GAME LOG CONTINUED

Date	Game Played	Players

Games help you understand, apply, and enjoy mathematics.

GAME LOG CONTINUED

Date	Game Played	Players

Games help you understand, apply, and enjoy mathematics.

Name: Jaxon

Date: 8/26/24

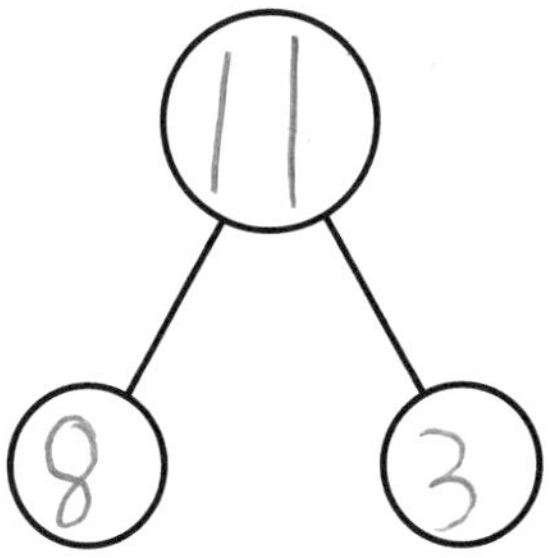

Eleven cherries were on the plate. Matthew ate some of them. Then there were eight cherries. How many cherries did Matthew eat?

11-3=8

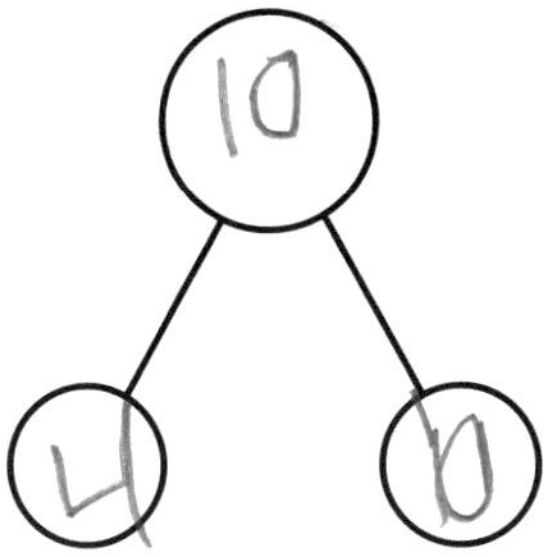

Four crows were sitting on the grass. More crows flew there. Then there were ten crows. How many crows flew over to the first four?

4+6=10

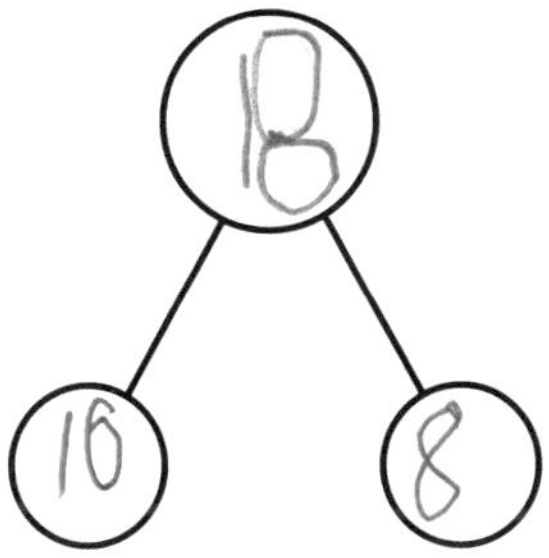

Elizabeth had ten goldfish. For her seventh birthday, she received eight goldfish. How many goldfish does she have now?

10+8=18

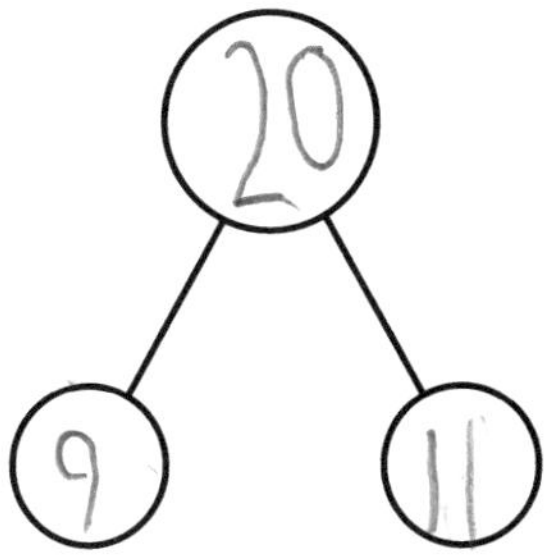

Maria picked nine apples. Jack picked two more apples than Maria picked. How many apples did Jack pick?

9+11=20

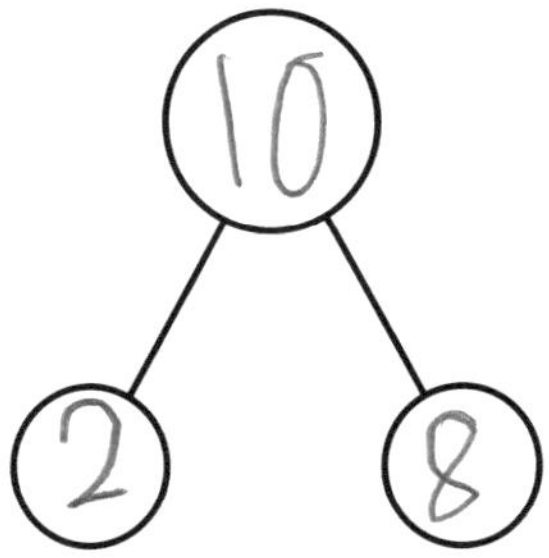

Morgan read 10 pages. Chris read two fewer pages. How many pages did Chris read?

10-2=8 or 2+8=10

Name: Jaxon Date: 8/29/24

Ones and Twos Strategies

+	1	2	3	4	5	6	7	8	9
1	2	3	4	5	6	7	8	9	10
2	3	4	5	6	7	8	9	10	11
3	4	5	6	7	8	9	10	11	12
4	5	6	7	8	9	10	11	12	13
5	6	7	8	9	10	11	12	13	14
6	7	8	9	10	11	12	13	14	15
7	8	9	10	11	12	13	14	15	16
8	9	10	11	12	13	14	15	16	17
9	10	11	12	13	14	15	16	17	18

Tens and Near Tens Strategies

+	1	2	3	4	5	6	7	8	9
1									
2									
3									
4									
5									
6									
7									
8									
9									

Two-Fives Strategy

+	1	2	3	4	5	6	7	8	9
1									
2									
3									
4									
5									
6									
7									
8									
9									

Doubles & Near Doubles Strategies

+	1	2	3	4	5	6	7	8	9
1									
2									
3									
4									
5									
6									
7									
8									
9									

Making Ten Strategy

+	1	2	3	4	5	6	7	8	9
1									
2									
3									
4									
5									
6									
7									
8									
9									

Addition Table

+	1	2	3	4	5	6	7	8	9
1									
2									
3									
4									
5									
6									
7									
8									
9									

Name: ______________________________

Use the Roman numeral chart to complete the following.

1	I
5	V
10	X
50	L
100	C
500	D
1000	M

16: ______ 44: ______

59: ______ 103: ______

290: ______ 377: ______

XXII: ______ LXVI: ______ LXXXXVII: ______

CXI: ______ CXX: ______ CLXXXII: ______

CC: ______ CCCXXXX: ______ CCCCXXXV: ______

Name: ____________________

Date: ____________________

1	I
5	V
10	X
50	L
100	C
500	D
1000	M

H-A	Early	Later
4	IIII	IV
9	VIIII	IX
40	XXXX	XL
90	LXXXX	XC

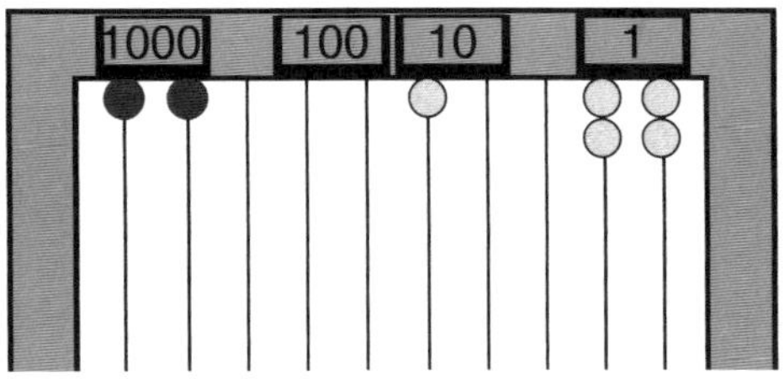

Fill in the blanks.

H-A	Early	Later
______	______	______

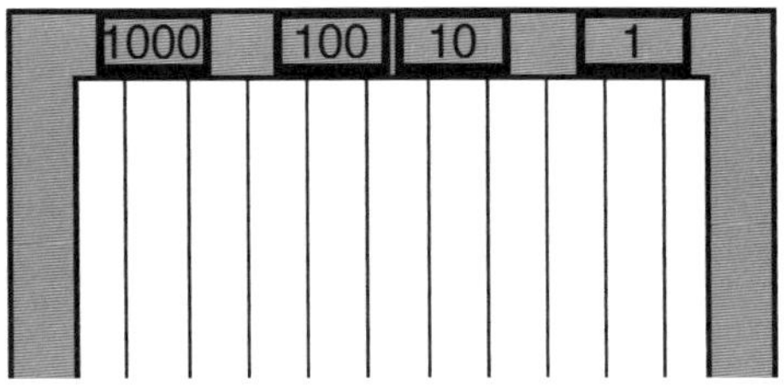

Draw the beads on the abacus and fill in the blanks.

H-A	Early	Later
3129	______	______

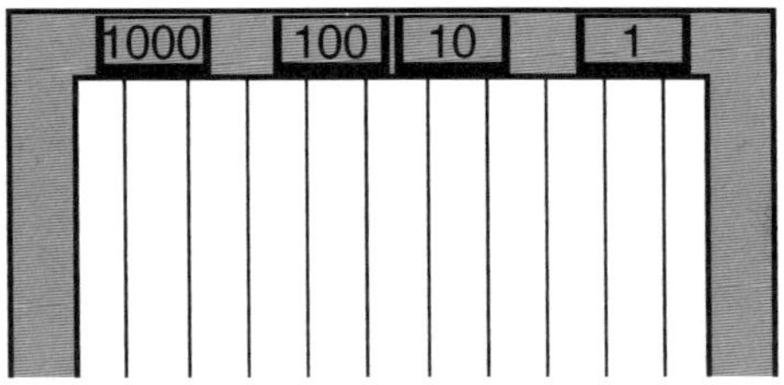

Draw the beads on the abacus and fill in the blanks.

H-A	Early	Later
______	DLXXII	______

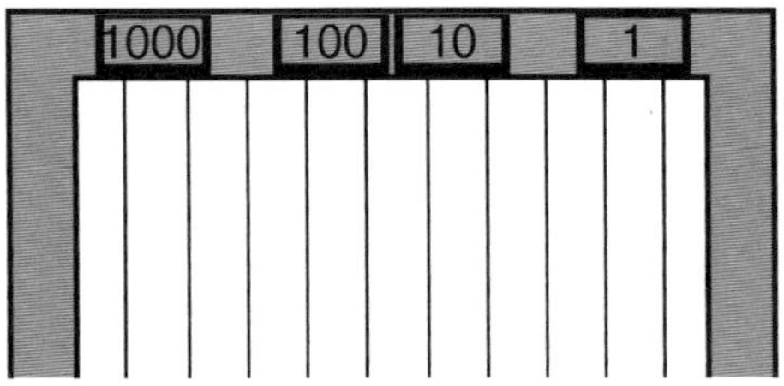

Draw the beads on the abacus and fill in the blanks.

H-A	Early	Later
______	______	MCCCXLVI

Name: Jaxon

Date: 8/27/24

Add.

65 + 16 = 81

29 + 43 = 72

34 + 58 = 92

55 + 55 = 110

79 + 62 = 141

99 + 87 = 186

18 + 15 = 33

59 + 61 = 120

47 + 67 = 114

73 + 27 = 100

98 + 37 = 135

88 + 88 = 176

Add 18 + 74 in two different ways and explain how you did it.

Name: Jaxon

Date: 8/29/24

Add.

35 + 26 + 9 + 31 = 101

2 + 73 + 98 + 6 = 173

16 + 12 + 2 + 93 = 123

86 + 2 + 29 + 22 = 139

13 + 1 + 37 + 15 = 66

54 + 37 + 8 + 25 = 124

71 + 52 + 70 + 32 = 225

23 + 70 + 53 + 6 = 152

Choose one of the equations and explain two different ways how to solve it.

____

____

____

____

____

____

Name: ____________________

Date: ____________________

Write only the answers.

______

______

______

Add.

38 + 6 = ______

99 + 64 = ______

36 + 14 + 50 + 47 = ______

Explain how to add 39 and 41 two different ways.

____

____

____

____

____

____

Write the number of squares with Hindu-Arabic numerals and with Roman numerals.

1	I
5	V
10	X

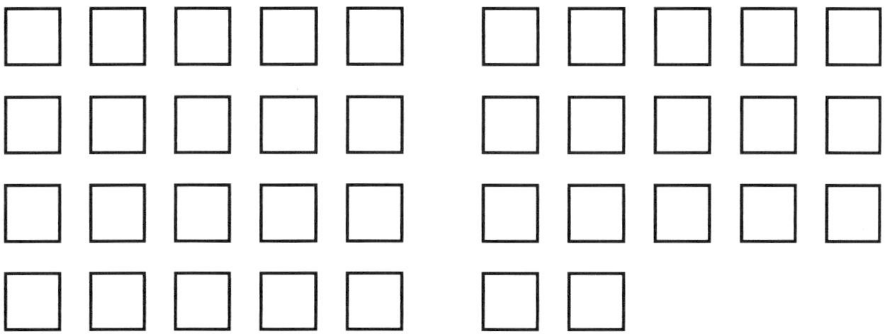

______

______

Name: ______________________________

Date: ______________________________

Write only the answers.

____

____

____

Add.

47 + 8 = ______

99 + 56 = ______

22 + 18 + 60 + 75 = ______

Explain how to add 21 and 19 two different ways.

____

____

____

____

____

____

Write the number of squares with Hindu-Arabic numerals and with Roman numerals.

1	I
5	V
10	X

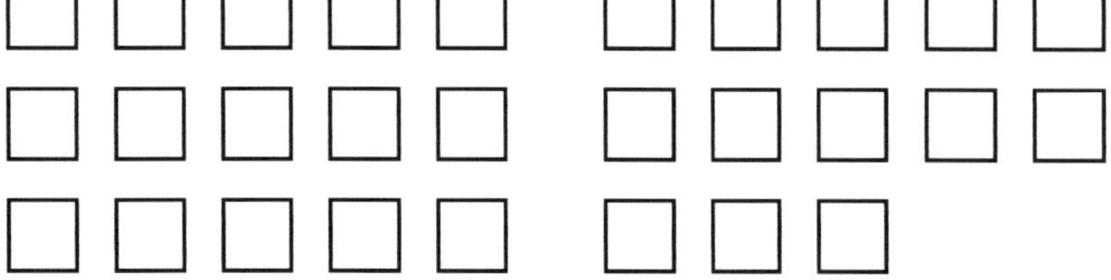

____

____

Name: ______________________________

Date: ______________________________

Write the expanded form of the circled base-10 cards. Draw beads on the abacus and write the number in standard form.

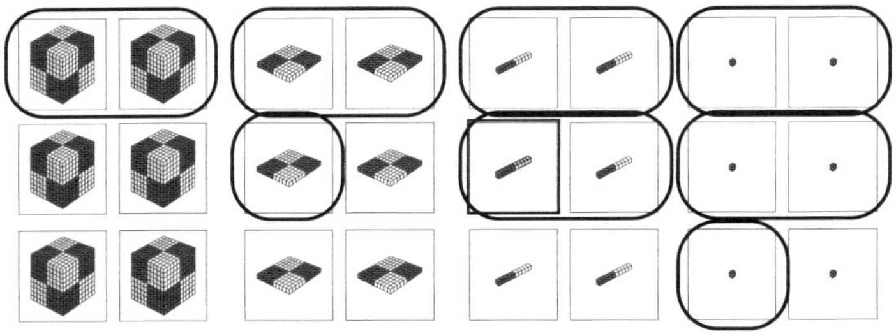

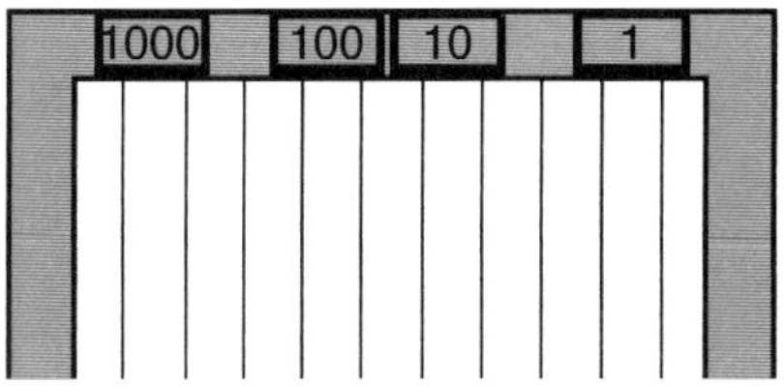

[_ 0 0 0] + [_ 0 0] + [_ 0] + [_]   [_ _ _ _]

Look at the number in expanded form. Circle the base-10 cards. Draw beads on the abacus and write the number in standard form.

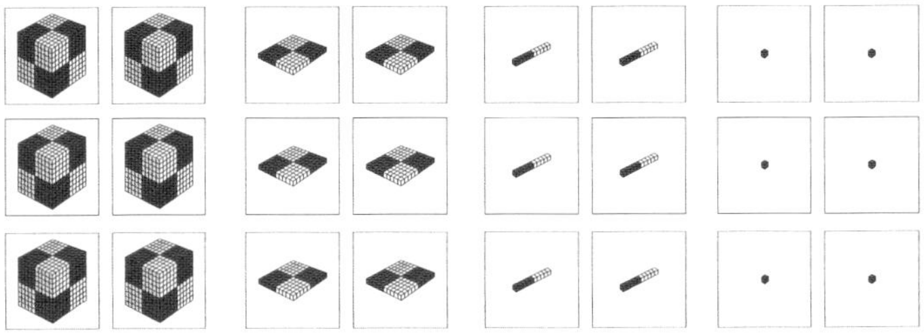

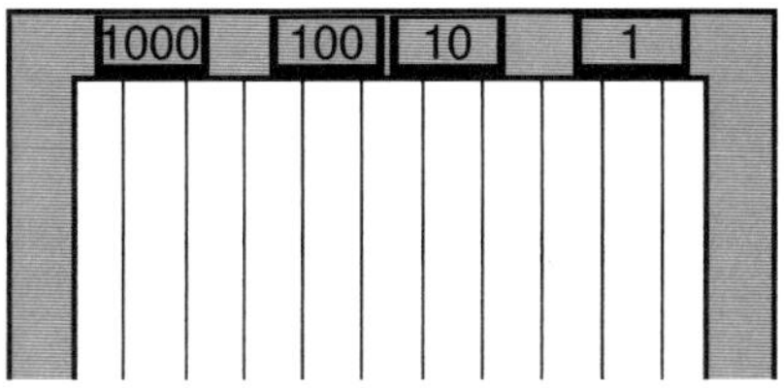

[1 0 0 0] + [2 0 0] + [6 0] + [4]   [_ _ _ _]

Look at the beads on the abacus. Circle the base-10 cards. Write the number expanded form and in standard form.

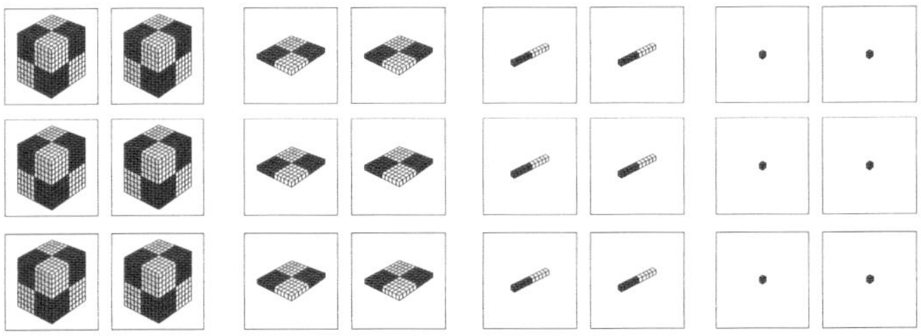

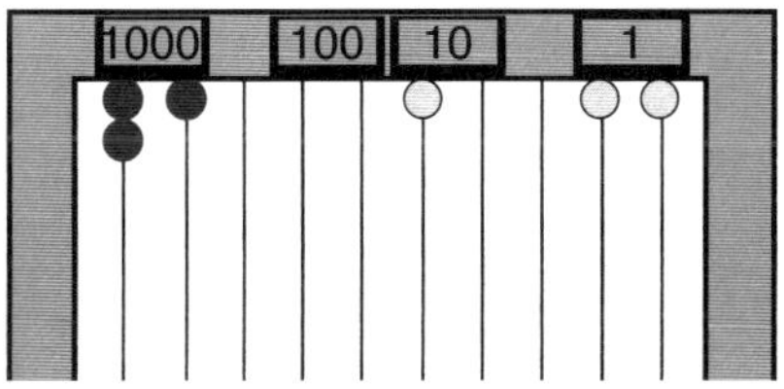

[_ 0 0 0] + [_ 0 0] + [_ 0] + [_]   [_ _ _ _]

Look at the number in standard form. Circle the base-10 cards. Write the number in expanded form and draw the beads on the abacus.

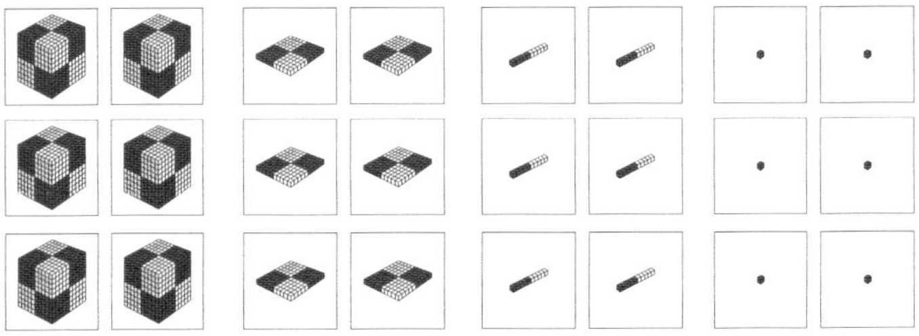

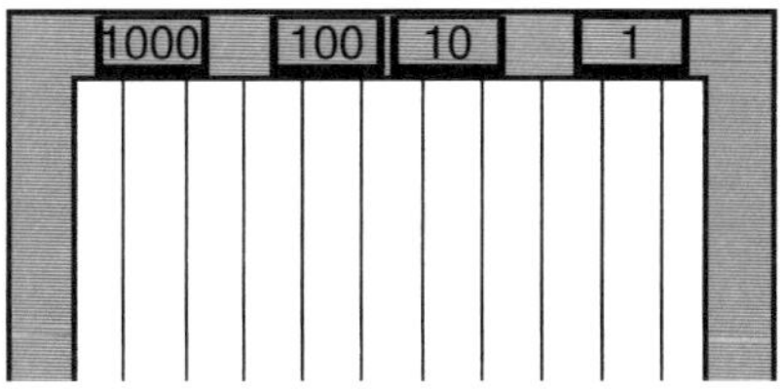

[4 2 0 6]

[_ 0 0 0] + [_ 0 0] + [_ 0] + [_]

Name: ______________________________

Date: ______________________________

Write >, <, or = on the lines to make the equations true.

38 + 6 _____ 30 + 6

506 _____ 560

99 + 10 _____ 109

250 + 10 _____ 251 + 9

700 + 80 _____ 708

1000 _____ 300 + 700

99 + 64 _____ 100 + 64

211 _____ 200 + 10

99 + 100 _____ 190

89 + 63 _____ 100 + 73

38 _____ 30 + 8

461 _____ 400 + 60

Write >, <, or = and explain your answer.

611 + 100 _____ 611 + 10

____

____

95 + 10 + 5 _____ 110

____

____

455 + 10 + 1 _____ 100 + 365

____

____

Name: ____________________

Date: ____________________

Solve the problems using the base-10 picture cards. Write the numbers in the grid. Explain how you got your answer.

1. Planners are building a swimming pool and need to know how many children live in the towns of Addie, Hammer, and Preston. Addie has 2697 children. Hammer has 3986 children and Preston has 1449 children.

2. Bird watchers counted 879 robins, 4387 finches, and 2718 swallows. How many birds did the watchers count?

Name: ______________________________

Date: ______________________________

	4	8	1	7
+	2	6	3	9

	3	6	2	9
+	2	5	8	4

	1	6	4	4
+	7	7	8	6

	2	4	8	2
+	5	6	1	3

	2	9	3	4
+	6	3	7	3

	6	8	7	9
+	4	2	1	7

Write the sum in words.

	4	2	5	8
+	3	1	6	6

____

	7	5	8	2
+		9	3	9

____

	6	0	5	8
+	6	2	7	3

____

	2	5	2	5
+	2	5	2	5

____

Name: ______________________________

Date: ______________________________

Write only the answers.

____

____

____

Add.

75 + 27 = ______

6 + 109 = ______

27 + 13 + 100 + 34 = ______

Write the number shown by the pictures.

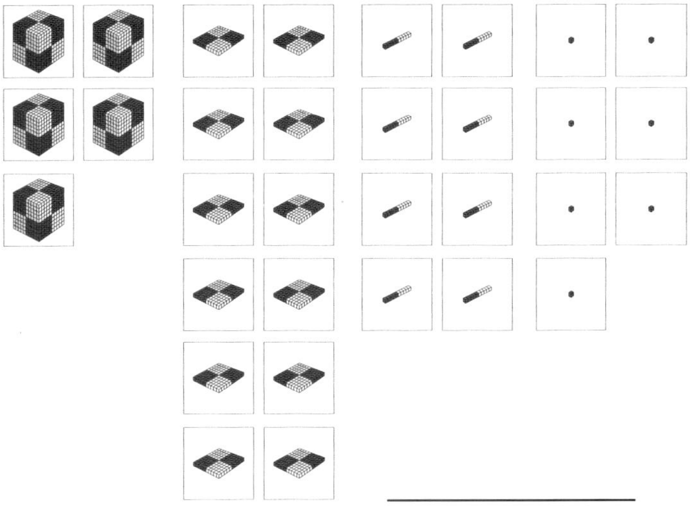

____

Add.

	3	8	4	7
+	5	0	8	5

Write >, <, or = on the lines.

138 _____ 30 + 80

101 + 11 _____ 122

Write this number in expanded form.

2017 ____

Write this number in standard form.

4 thousand 8 hundred thirty-seven ___________

Name: ______________________________

Date: ______________________________

Write only the answers.

____

____

____

Add.

35 + 88 = ______

7 + 106 = ______

42 + 18 + 100 + 24 = ______

Write the number shown by the pictures.

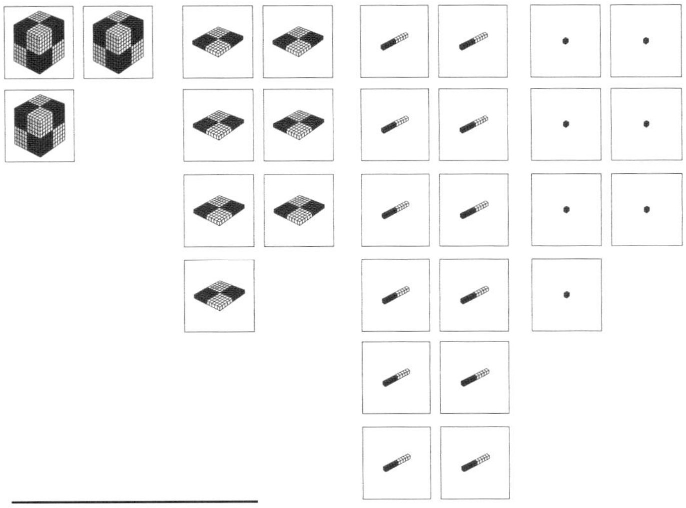

____

Add.

	4	9	3	7
+	3	5	4	8

Write >, <, or = on the lines.

127 _____ 40 + 90

101 + 16 _____ 117

Write this number in expanded form.

3560 ____

Write this number in standard form.

3 thousand 9 hundred twenty-four ____________

Name: ______________________________

Date: ______________________________

Write the array name and write one of the equations.

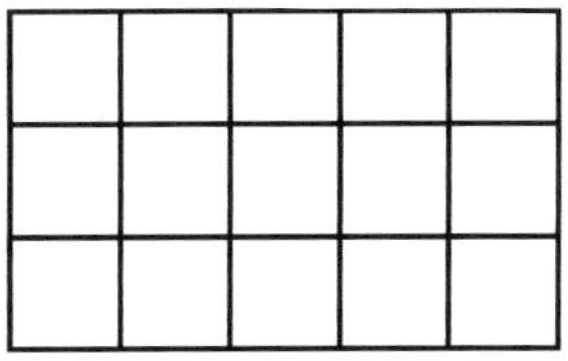

Array name ____________

Equation ______________________________

Draw the array and write one of the equations.

Array name 2 by 4

Equation ______________________________

Draw the array and write the array name.

Array name ____________

Equation 4 + 4 + 4 + 4 = ____

Draw the array and write one of the equations.

Array name 3 by 2

Equation ______________________________

Write the array name and write one of the equations.

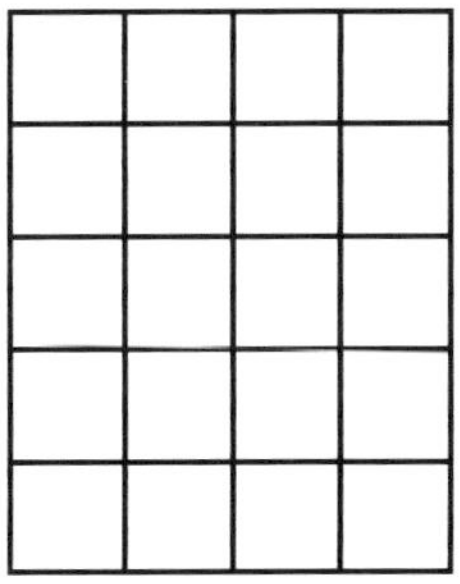

Array name ____________

Equation ______________________________

Name: ______________________________

Date: ______________________________

Find the amount in each array. Use your abacus when it helps.

$5 \times 2 =$ ______

$4 \times 10 =$ ______

$5 \times 5 =$ ______

$2 \times 7 =$ ______

$5 \times 10 =$ ______

$7 \times 7 =$ ______

$3 \times 4 =$ ______

$8 \times 2 =$ ______

$9 \times 3 =$ ______

$6 \times 4 =$ ______

$3 \times 8 =$ ______

$7 \times 3 =$ ______

$10 \times 10 =$ ______

$4 \times 4 =$ ______

Write >, <, or = on the lines. Use your abacus when it helps.

2×1 ____ $2 + 1$

2×2 ____ $2 + 2$

1×4 ____ $1 + 4$

5×1 ____ $5 + 1 - 1$

3×4 ____ $4 + 4 + 4$

8×1 ____ $8 + 1$

2×3 ____ $2 + 3$

2×4 ____ $2 + 1 + 5$

3×3 ____ $3 + 3 + 3$

10×10 ____ $10 + 10$

Name: Jaxon Date: 11/12/24

Multiples of 2.	Multiples of 5.	Multiples of 3.	Multiples of 4.
2 × 1 = 2	5 × 1 = 5	3 × 1 = 3	4 × 1 = 4
2 × 2 = 4	5 × 2 = 10	3 × 2 = 6	4 × 2 = 8
2 × 3 = 6	5 × 3 = 15	3 × 3 = 9	4 × 3 = 12
2 × 4 = 8	5 × 4 = 20	3 × 4 = 12	4 × 4 = 16
2 × 5 = 10	5 × 5 = 25	3 × 5 = 15	4 × 5 = 20
2 × 6 = 12	5 × 6 = 30	3 × 6 = 18	4 × 6 = 24
2 × 7 = 14	5 × 7 = 35	3 × 7 = 21	4 × 7 = 28
2 × 8 = 16	5 × 8 = 40	3 × 8 = 24	4 × 8 = 32
2 × 9 = 18	5 × 9 = 45	3 × 9 = 27	4 × 9 = 36
2 × 10 = 20	5 × 10 = 50	3 × 10 = 30	4 × 10 = 40

Name: ______________________________

Date: ______________________________

1. Which rectangle takes up more space? Explain.

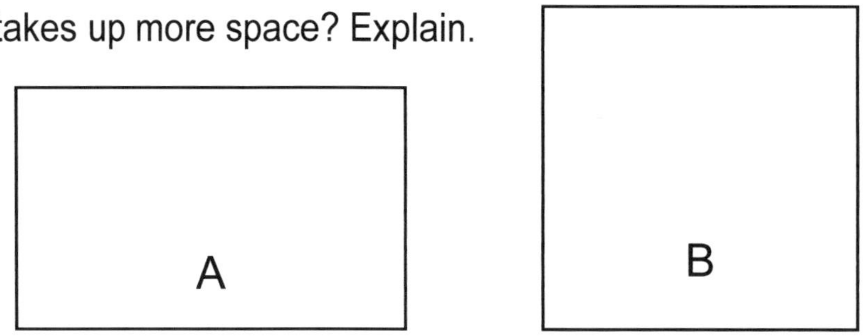

____

____

2. Ladybugs crawled around the sides of rectangle A and rectangle B. Which ladybug crawled farther? Explain.

____

____

Find the area of each rectangle and the distance around it.

3. 

____

____

4. 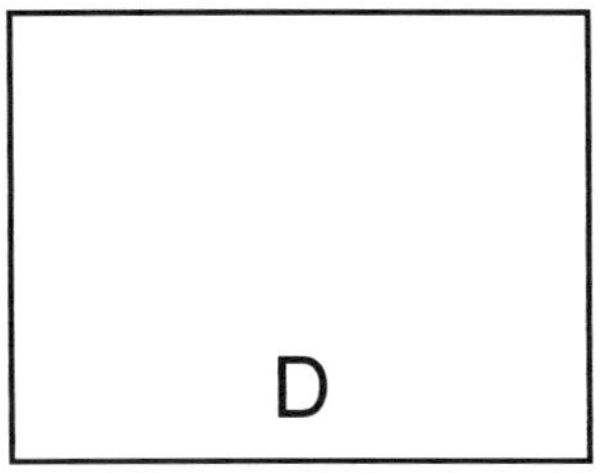

____

____

5. 

____

____

Name: ______________________________

Date: ______________________________

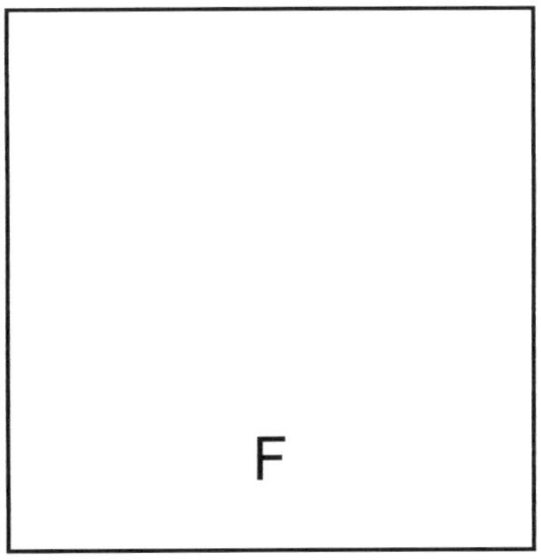

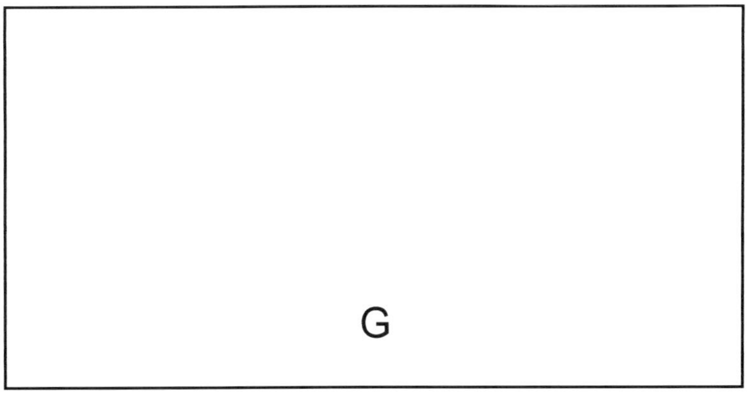

1. Find the perimeter of rectangle F with tiles.

____

1 inch

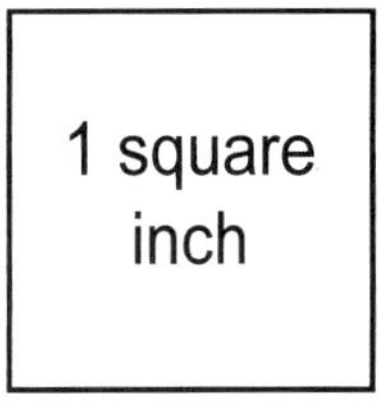

2. Find the perimeter of rectangle G with tiles.

____

3. Find the area of rectangle F with tiles.

____

4. Find the area of rectangle G with tiles.

____

5. Find the perimeter of rectangle F with centimeter cubes.

____

1 cm

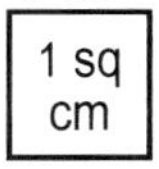

6. Find the perimeter of rectangle G with centimeter cubes.

____

7. Find the area of rectangle F with centimeter cubes.

____

8. Find the area of rectangle G with centimeter cubes.

____

Name: ______________________________

Date: ______________________________

Write only the answers.

______

87 + 25 + 3 = ______

______

5 by 2 = ______

______

6 × 1 = ______

	4	5	2	9
+	5	3	8	5

Find the perimeter in cm and the area in square cm.

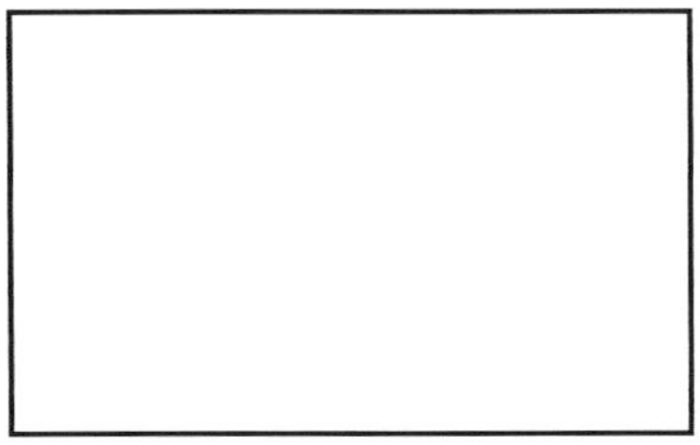

____

____

Fill in the missing place-value card numbers.

_ 0 0 0 + _ 0 0 + _ 0 + _   _ _ _ _

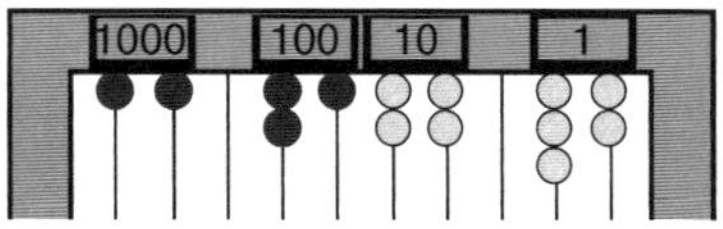

_ 0 0 0 + _ 0 0 + _ 0 + _   _ _ _ _

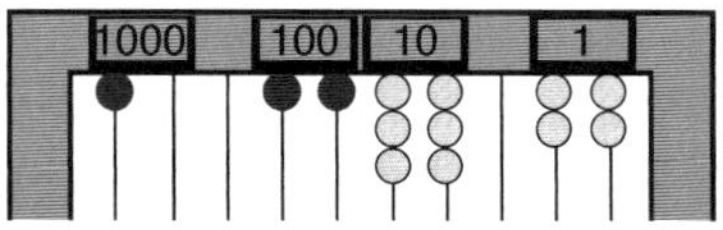

Write >, <, or = on the lines.

3 × 1 _____ 3 + 1

1 inch _____ 1 cm

Draw a 4 × 3 array. How much is it?

______

Write this number in standard form.

seven thousand one hundred thirteen ____________

Write this number in words.

4807 ____

Name: ______________________________

Date: ______________________________

Write only the answers.

__________ 79 + 37 + 1 = ______

__________ 5 by 3 = ______

__________ 1 × 4 = ______

	3	5	3	8
+	4	7	7	5

Find the perimeter in cm and the area in square cm.

____

____

Fill in the missing place-value card numbers.

_000 + _00 + _0 + _   _ _ _ _

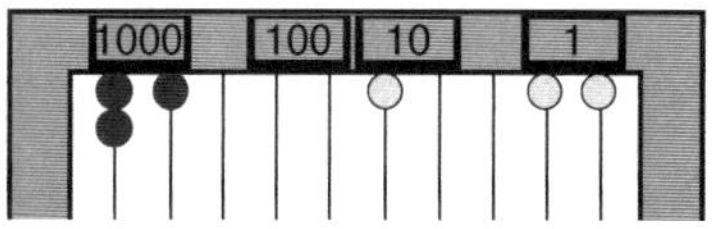

_000 + _00 + _0 + _   _ _ _ _

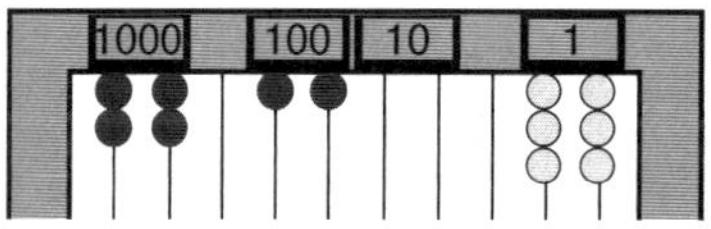

Write >, <, or = on the lines.

2 × 2 _____ 2 + 2

1 cm _____ 1 inch

Draw a 3 × 4 array. How much is it?

____

Write this number in standard form.

eight thousand four hundred fifteen ___________

Write this number in words.

3260 ____

Name: ______________________________

Date: ______________________________

Solve the problems using the part-whole circles. Write the equations.

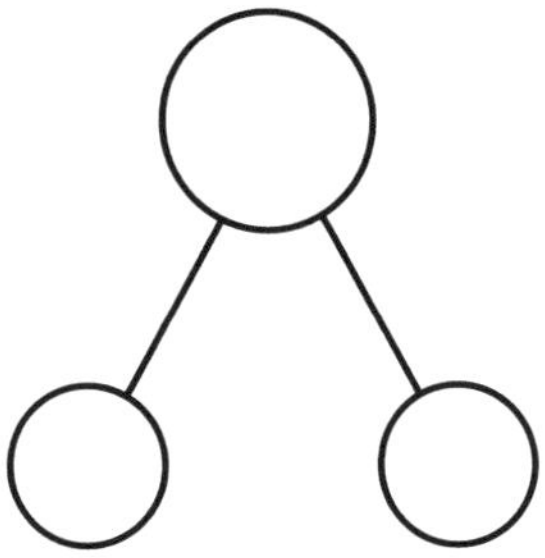

1. There are 17 boys and 14 girls in a chorus. How many children are in the chorus?

____

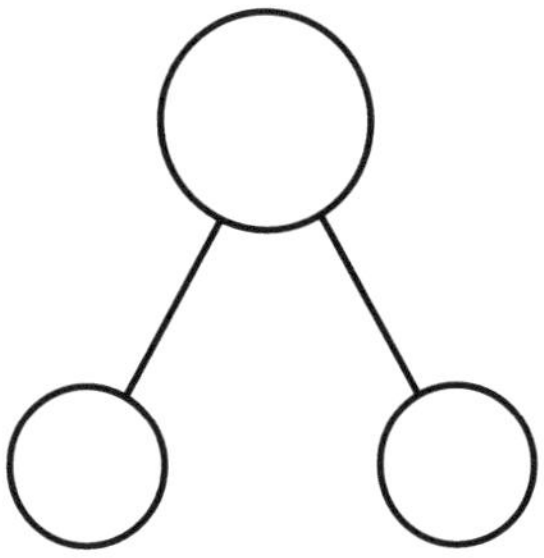

2. Jack wants to buy a gift that costs one dollar. A dollar is 100 cents. Jack already has 85¢. How much more money does he need?

____

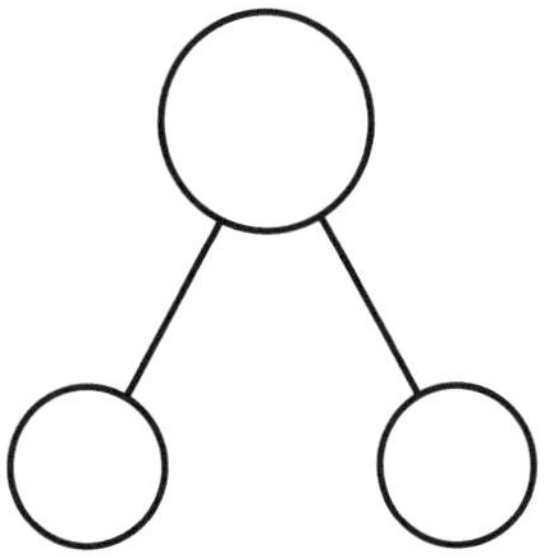

3. Jill is packing 24 bagels in a bag. She has 9 bagels in the bag already. How many more bagels must she put into the bag?

____

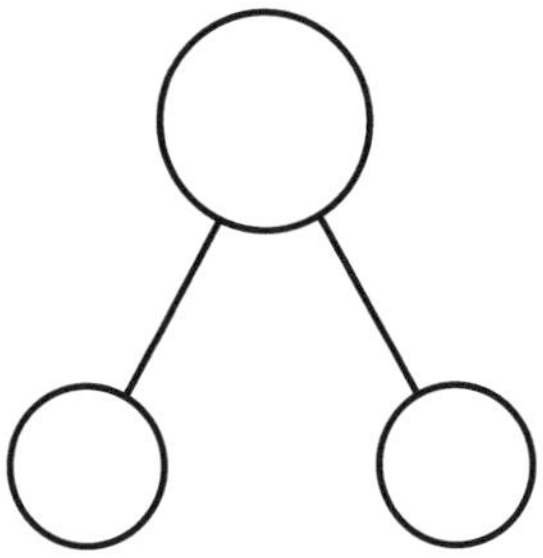

4. On Monday Kerry read 11 pages in a book. Now Kerry is on page 34. What page did Kerry start on?

____

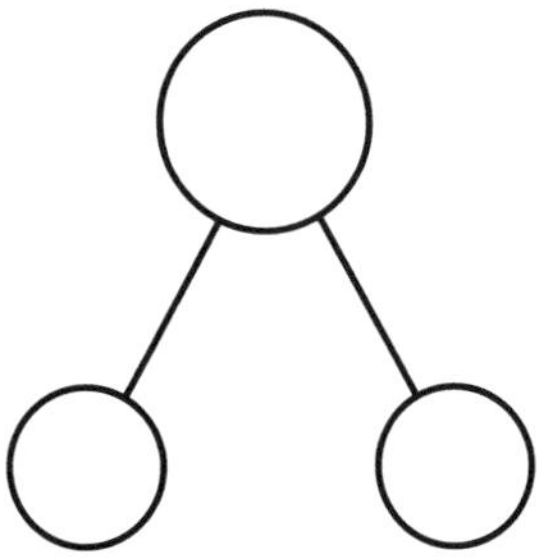

5. Some children were lined up. Fourteen more children join the line. Now there are 21 children. How many children were in line at the start?

____

Name: ______________________________

Date: ______________________________

Ones and Twos Strategies

	1	2	3	4	5	6	7	8	9	10	11	12	13	14	15	16	17	18
–1			2															
–2																		
–3																		
–4																		
–5																		
–6																		
–7																		
–8																		
–9																		

Consecutive and Same Numbers Strategies

	1	2	3	4	5	6	7	8	9	10	11	12	13	14	15	16	17	18
–1																		
–2																		
–3																		
–4																		
–5																		
–6																		
–7																		
–8																		
–9																		

Name: ______________________________

Date: ______________________________

Tens and Near Tens Strategies

	1	2	3	4	5	6	7	8	9	10	11	12	13	14	15	16	17	18
–1																		
–2																		
–3																		
–4																		
–5																		
–6																		
–7																		
–8																		
–9																		

Take from Five Strategies

	1	2	3	4	5	6	7	8	9	10	11	12	13	14	15	16	17	18
–1																		
–2																		
–3																		
–4																		
–5																		
–6																		
–7																		
–8																		
–9																		

Name: ______________________________

Date: ______________________________

Subtracting from Ten Strategies

	1	2	3	4	5	6	7	8	9	10	11	12	13	14	15	16	17	18
–1																		
–2																		
–3																		
–4																		
–5																		
–6																		
–7																		
–8																		
–9																		

Subtraction Table

	1	2	3	4	5	6	7	8	9	10	11	12	13	14	15	16	17	18
–1																		
–2																		
–3																		
–4																		
–5																		
–6																		
–7																		
–8																		
–9																		

Name: ____________________

Date: ____________________

Write only the answers.

____

____

____

Write the answers.

67 + 25 – 7 = ______

4 by 3 = ______

13 – 4 = ______

Find the perimeter in cm and the area in square cm.

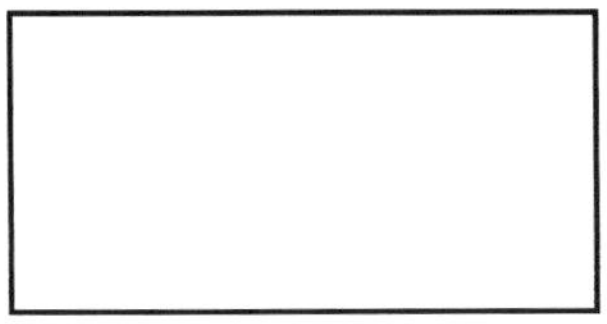

____

____

Add.

	5	6	8	3
+	3	3	5	6

Fill in the subtraction table.

	9	10	11	12
–4				
–5				
–6				
–7				

Write >, <, or = on the lines.

4 – 1 _____ 4 + 1

9 – 6 _____ 10 – 7

13 – 5 _____ 13 – 6

3 inches _____ 3 cm

8 – 6 _____ 7 – 5

9 – 8 _____ 0 + 6

Explain two ways how to find 14 – 5.

____

____

Name: ______________________________

Date: ______________________________

Write only the answers.

____

____

____

Write the answers.

74 + 15 – 4 = ______

3 by 4 = ______

16 – 7 = ______

Find the perimeter in cm and the area in square cm.

____

____

Add.

	4	2	7	3
+	2	7	8	7

Fill in the subtraction table.

	10	**11**	**12**	**13**
–5				
–6				
–7				
–8				

Write >, <, or = on the lines.

5 + 2 _____ 5 – 2

8 – 5 _____ 11 – 8

14 – 6 _____ 14 – 5

2 inches _____ 2 cm

10 – 8 _____ 5 – 3

9 + 7 _____ 8 – 0

Explain two ways how to find 11 – 6.

____

____

Name: ____________________

Date: ____________________

Complete the subtraction equations using only these numbers:

18, 20, 21, 24, 27, 30, 32, 35, 40, 45, 49, 50, 56, 60, and 64.

1 = 21 – 20	21 = ______
2 = ______	22 = ______
3 = ______	23 = ______
4 = ______	24 = ______
5 = ______	25 = ______
6 = ______	26 = ______
7 = ______	27 = ______
8 = ______	28 = ______
9 = ______	29 = ______
10 = ______	30 = ______
11 = ______	31 = ______
12 = ______	32 = ______
13 = ______	33 = ______
14 = ______	34 = ______
15 = ______	35 = ______
16 = ______	36 = ______
17 = ______	37 = ______
18 = ______	38 = ______
19 = ______	39 = ______
20 = ______	40 = ______

Name: ______________________

Date: ______________________

Fix the error three different ways.

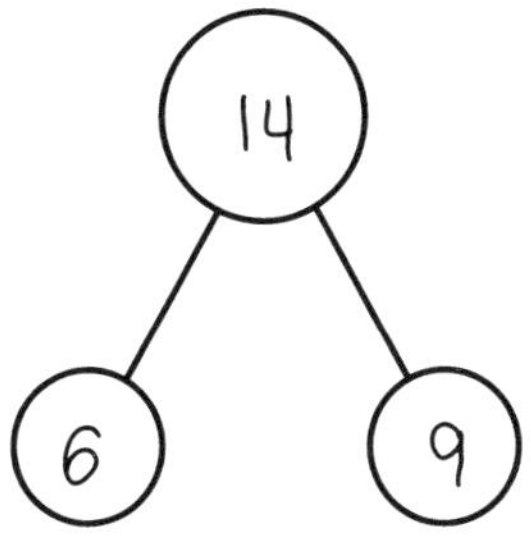

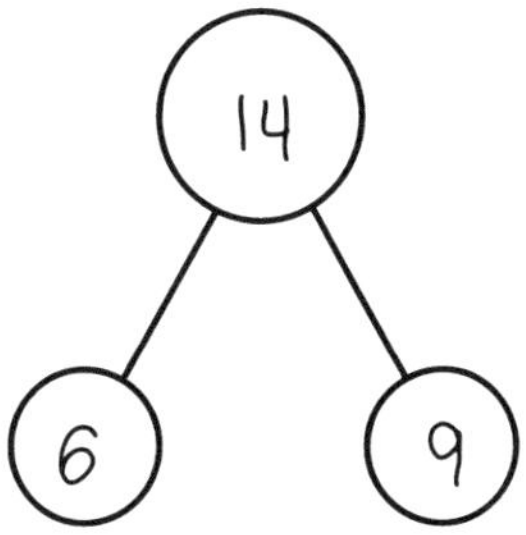

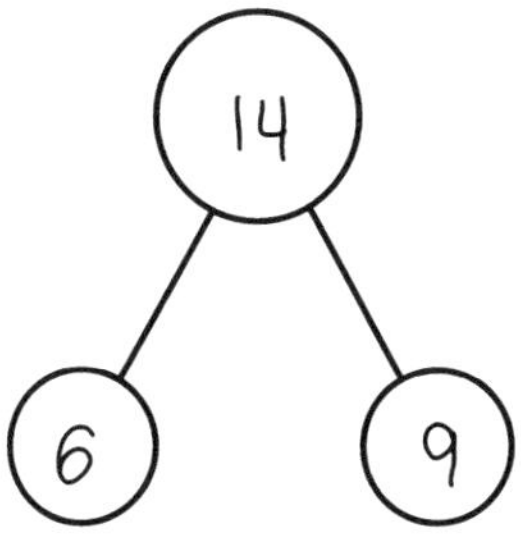

Fix the three errors in each table.

	11	12	13	14	15
–6	5	6	7	8	9
–7	4	5	8	7	8
–8	3	4	5	6	8
–9	2	4	4	5	6

+	4	5	6	7	8
6	9	10	11	12	13
6	10	11	12	13	14
7	11	12	13	14	1
8	12	13	4	15	16

Fix the array name.

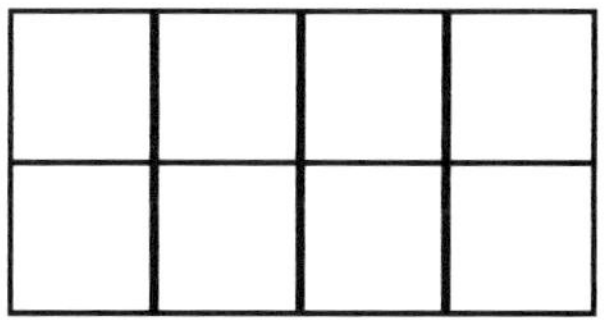

5 by 3 ______

Explain the errors; do not fix them.

	5	6	4	3
+	2	7	9	8
	7	3	3	1

____

	1	1	1	
	2	5	2	6
+	2	3	3	3
	5	9	6	9

____

$9 - 7 =$ 16

$9 + 7 =$ 2

____

Correct the errors.

five thousand fifteen 500015 ______

3206 three thousand twenty-six ______

2000 + 400 + 10 241 ______

Name: ______________________________

Date: ______________________________

Subtract.

68 – 42 = ______

170 – 134 = ______

87 – 39 = ______

141 – 36 = ______

93 – 79 = ______

122 – 96 = ______

193 – 114 = ______

66 – 42 = ______

56 – 31 = ______

105 – 69 = ______

100 – 47 = ______

100 – 18 = ______

100 – 85 = ______

100 – 36 = ______

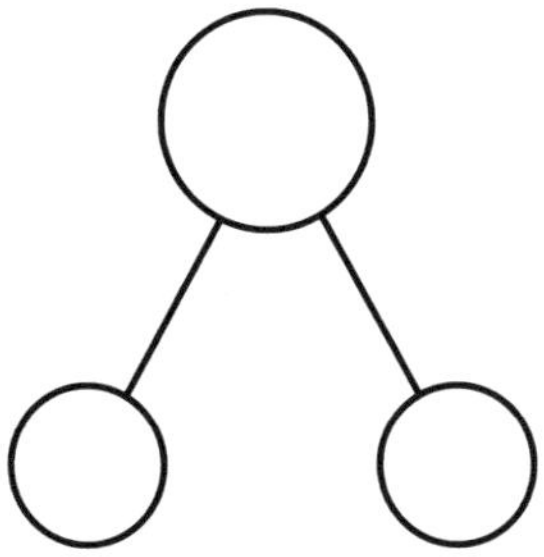

There are 123 animals altogether at a zoo. The zoo workers still have 87 animals to feed. How many animals have they already fed?

____

____

Explain two ways how to find 53 – 24.

____

____

Name: ______________________________

Date: ______________________________

Write only the answers.

____

____

____

Write the answers.

67 + 25 + 13 = ______

5 by 4 = ______

47 – 9 = ______

Find the perimeter in cm and the area in square cm.

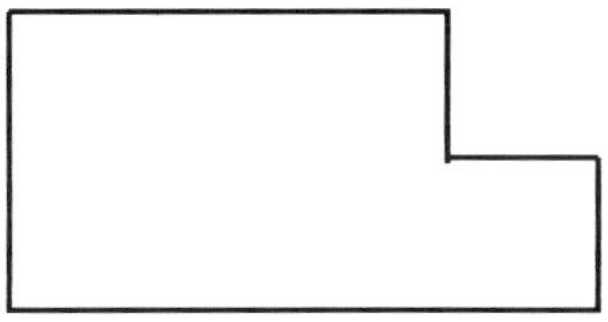

____

____

Fix the answer.

	4	7	5	2
+	1	9	8	5
	6	6	3	7

Find the differences.

42 – 28 = ______

100 – 63 = ______

123 – 49 = ______

Write >, <, or = on the lines.

10 – 2 _____ 11 – 3

9 – 6 _____ 10 – 6

15 – 6 _____ 15 – 7

1 cm _____ 2 cm

18 + 6 _____ 19 + 5

9 – 8 _____ 9 – 1

Explain two ways how to find 87 – 39.

____

____

Name: ______________________________

Date: ______________________________

Write only the answers.

____

____

____

Write the answers.

74 + 15 + 16 = ______

3 by 5 = ______

56 – 7 = ______

Find the perimeter in cm and the area in square cm.

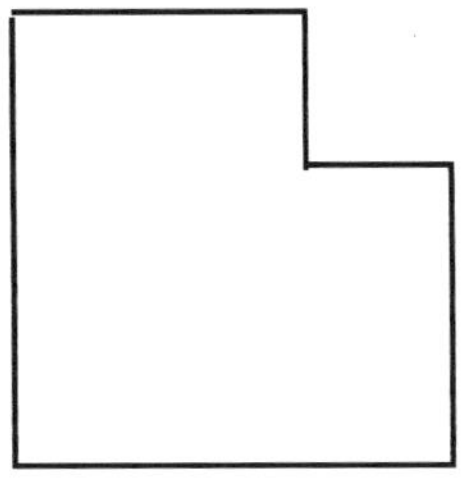

____

____

Fix the answer.

	4	2	7	3
+	2	7	8	7
	6	0	6	0

Find the differences.

63 – 25 = ______

100 – 74 = ______

137 – 59 = ______

Write >, <, or = on the lines.

5 – 2 _____ 7 – 4

8 – 7 _____ 11 – 7

14 – 6 _____ 14 – 5

2 cm _____ 3 cm

13 – 8 _____ 13 – 9

9 + 17 _____ 8 + 18

Explain two ways how to find 46 – 18.

____

____

Name: ______________________ Date: ______________________

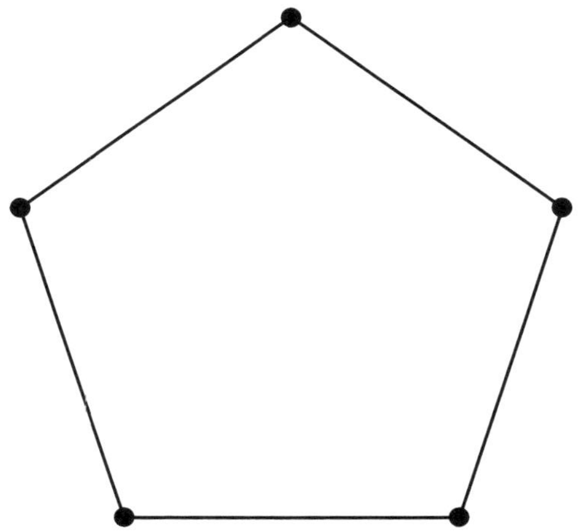

Draw the diagonals in this pentagon.

How many diagonals are there? ______

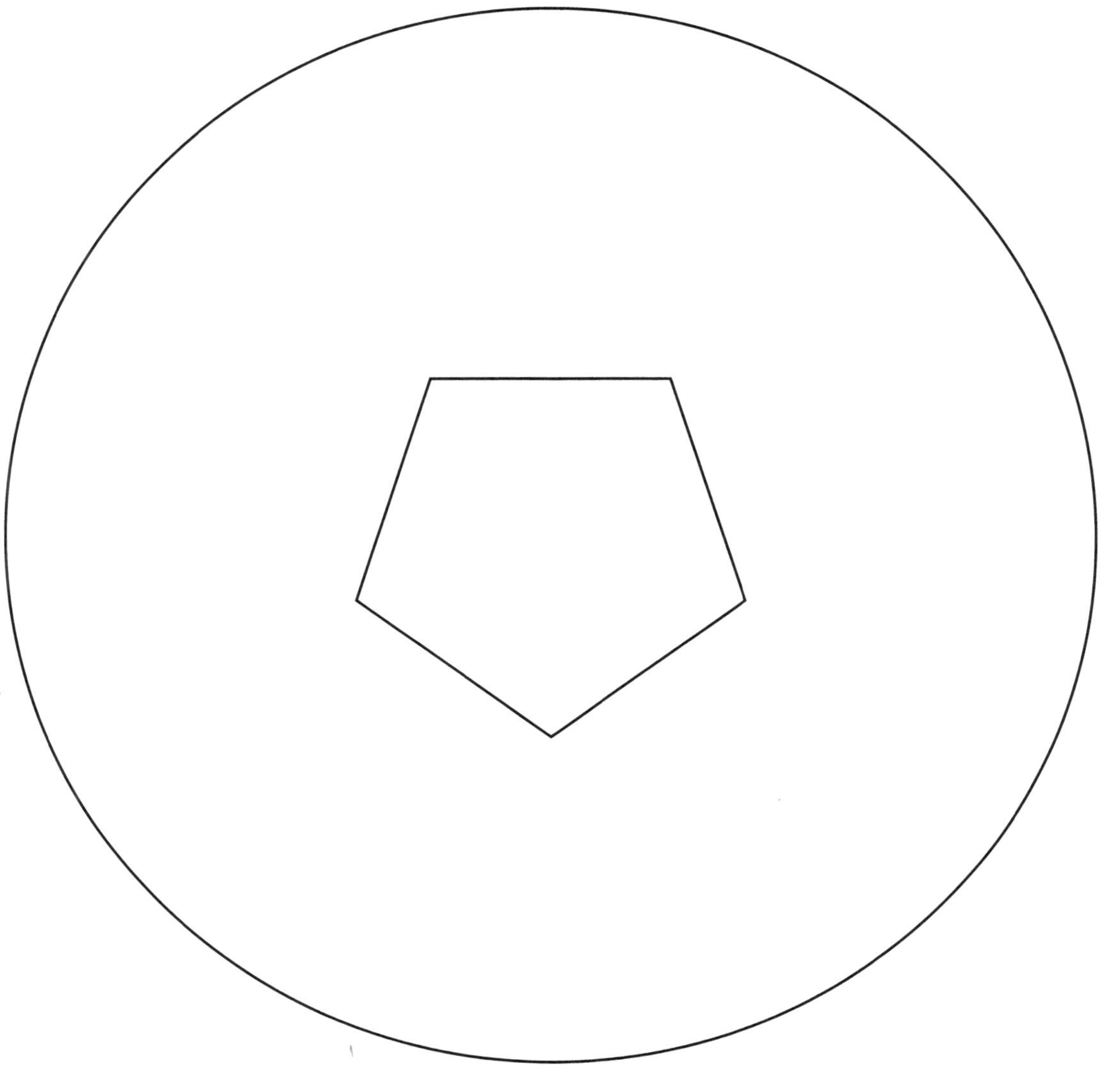

Extend the sides of this pentagon until they touch the circle.

Name: ______________________ Date: ______________________

Use your drawing tools to draw the lines.

Draw the rungs of the ladder.

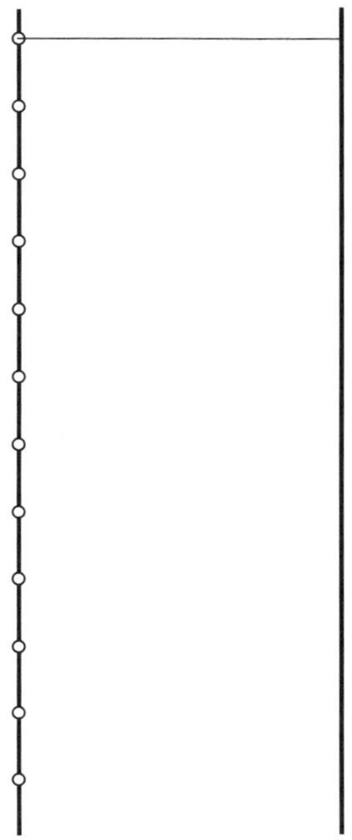

Complete the three rectangles.

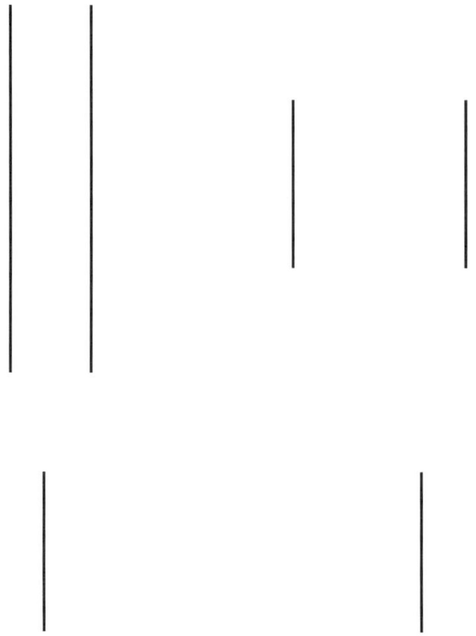

Draw the horizontal diagonals.

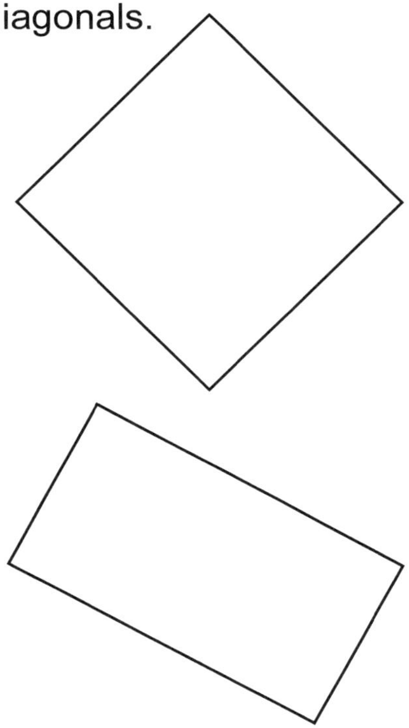

Divide these figures in half and write the fraction, $\frac{1}{2}$, in each of the parts.

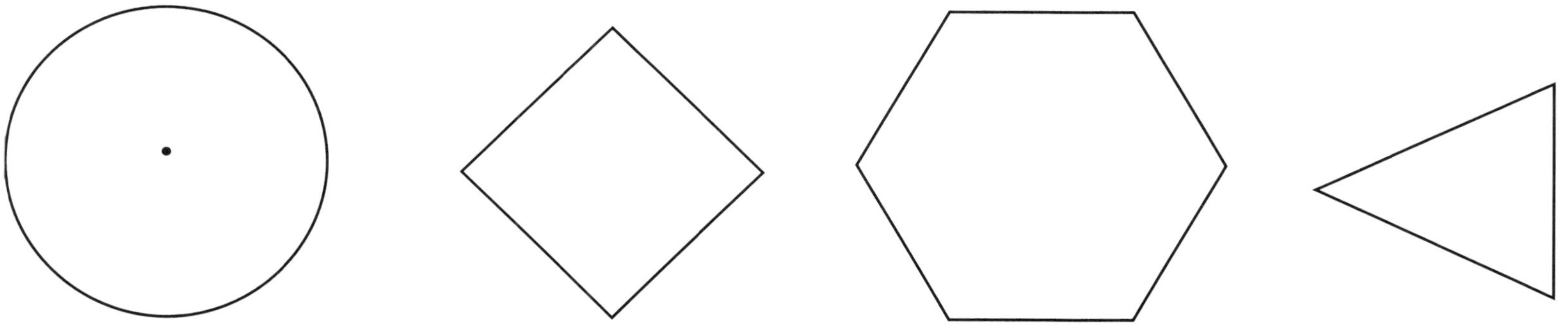

Name: ______________________ Date: ______________________

Use your drawing tools to draw all lines.

Draw the slats for the fence.

Complete the octagon.

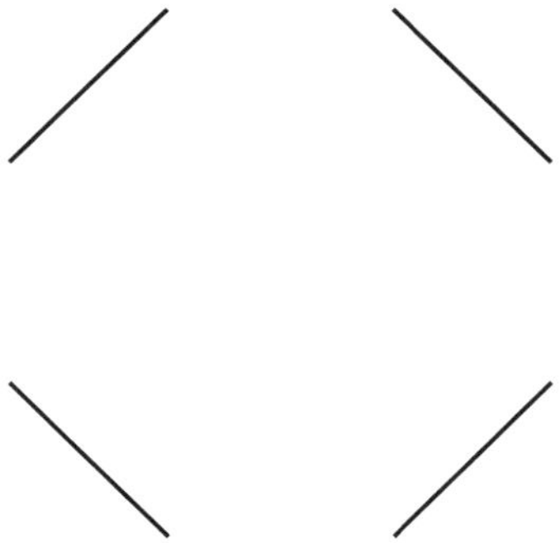

Draw both diagonals in the rhombuses.

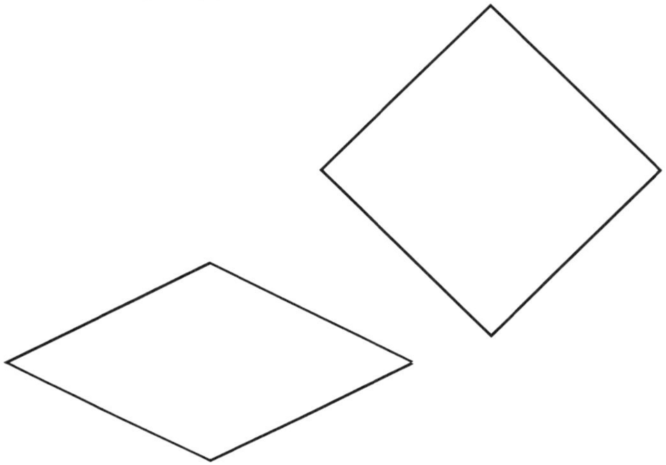

Divide each figure into four equal parts and write the fraction, $\frac{1}{4}$, in one of the parts.

Are the four parts in each figure congruent? _____

Name: ____________________ Date: ____________________

Use drawing tools to draw all the diagonals in the hexagon.

There will be three diagonals at each vertex.

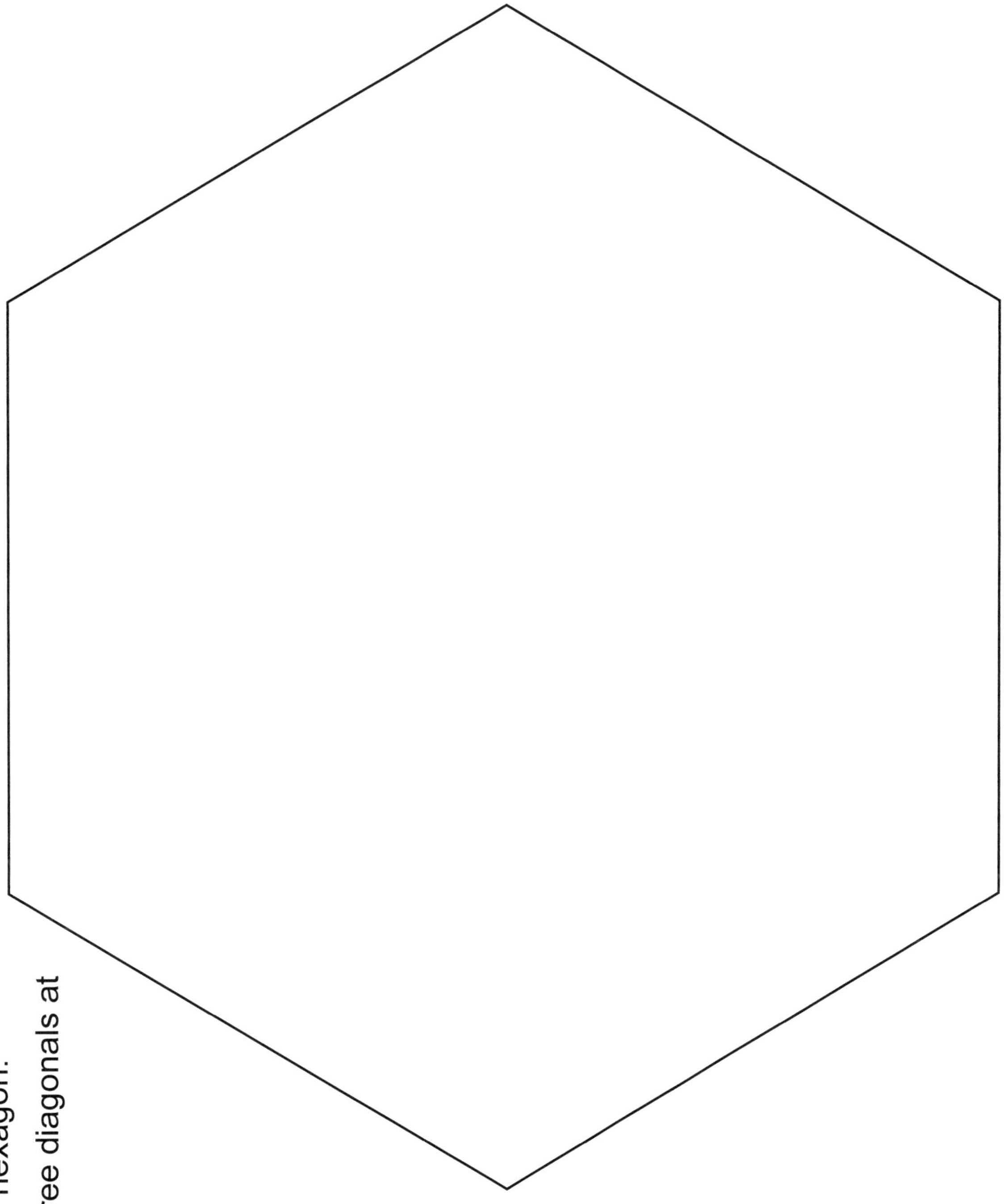

Name:______________________ Date: ______________________

Use your drawing tools to draw all lines.

1. Draw an equilateral triangle on each line. Divide each triangle in half a different way.

2. Draw three rows of equilateral triangles as shown.

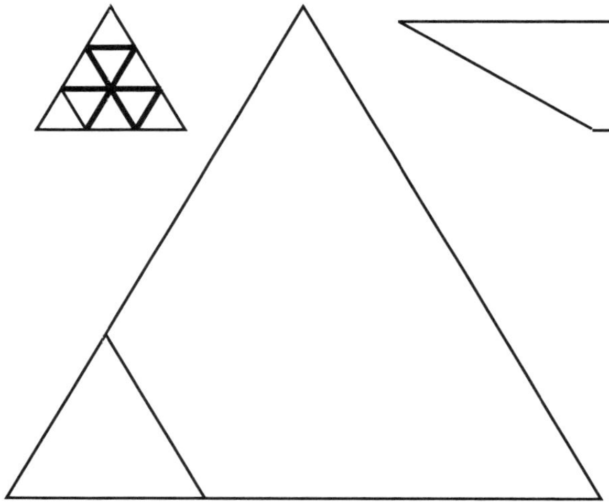

3. Draw zigzags as shown.

How many triangles are there altogether? _____

Name:______________________ Date: ______________________

Use your drawing tools to draw all lines.

Draw an equilateral triangle on each line. Then divide each triangle into thirds as shown.

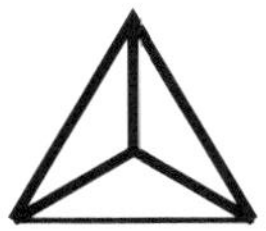

Divide into thirds.

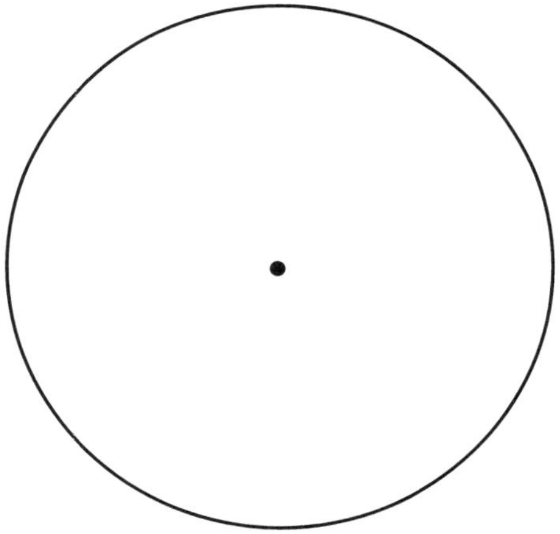

Divide into thirds.

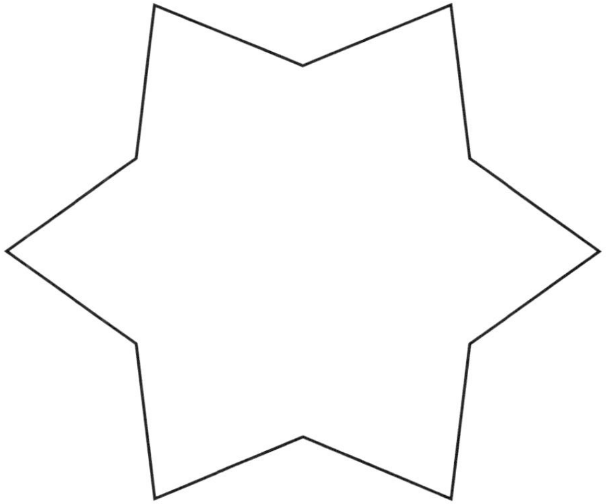

Circle $\frac{1}{3}$ of the triangles.

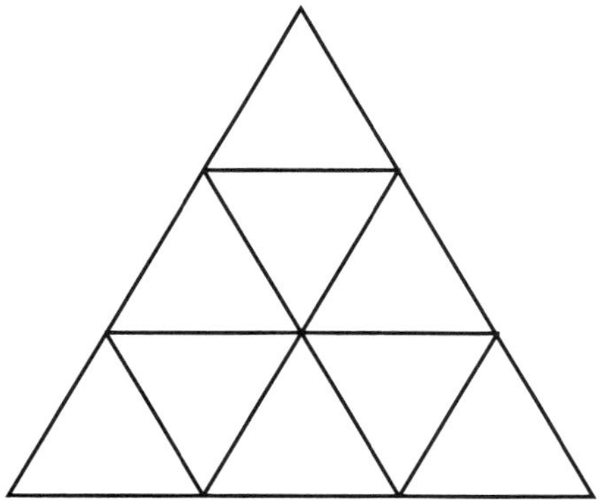

Name: ______________________ Date: ______________________

Use your drawing tools to draw all lines.

Draw equilateral triangles on both lines and divide them into fourths as shown.

1. ______________________

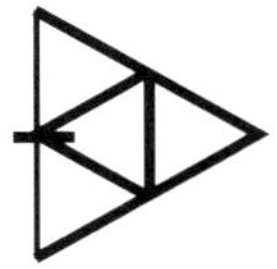

2. ______________________

Draw an equilateral triangle on the line and divide it into eighths.

3. ______________________

Complete the fraction chart.

<table>
<tr><td colspan="8">__</td></tr>
<tr><td colspan="4">__</td><td colspan="4">$\frac{1}{2}$</td></tr>
<tr><td colspan="2">__</td><td colspan="2">$\frac{1}{4}$</td><td colspan="2">__</td><td colspan="2">__</td></tr>
<tr><td>__</td><td>__</td><td>$\frac{1}{8}$</td><td>__</td><td>$\frac{1}{8}$</td><td>__</td><td>__</td><td>__</td></tr>
</table>

Name: ____________________ Date: ____________________

1. Draw an equilateral triangle on the line.
   Divide it into fourths.

Figure 1.

3. Cut out Figure 1 and fold it on the lines to make a tetrahedron.
4. Cut out Figure 2 and fold it on the sides of the square to make a square pyramid.

5. Complete the table.

	Faces	Edges	Vertices
Tetrahedron			
Square Pyramid			

2. Draw an equilateral triangle on each side of the square as shown.

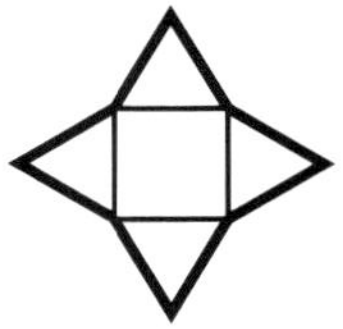

Figure 2.

Name: ______________________________

Date: ______________________________

Measure the lines in cm.

___ ____________ _______ _____

Draw an equilateral triangle on the line. Divide it into fourths. Then divide each fourth into thirds as shown.

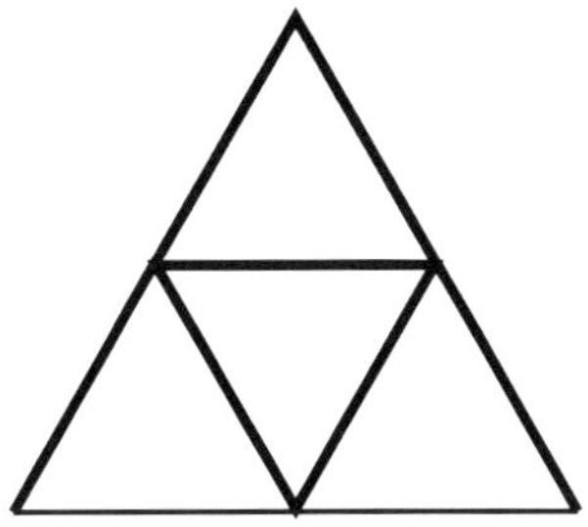

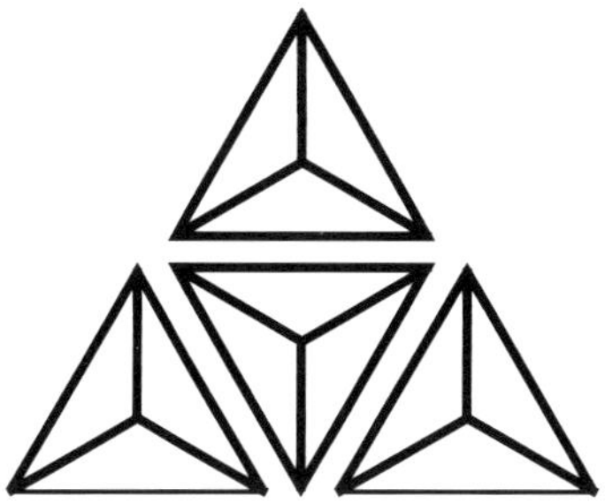

____

Find the perimeter in cm.

Large equilateral triangle *P* = ______________________ *P* = ______

Small equilateral triangle *P* = ______________________ *P* = ______

Small rhombus *P* = ______________________ *P* = ______

Hexagon *P* = ______________________ *P* = ______

Name: ______________________ Date: ______________________

Draw equilateral triangles on the lines using the drawing tools.

______________________ ______________________ ______________________

Divide the triangle into thirds. Color one third.

Divide the triangle into thirds a different way. Color a third.

Divide the triangle into sixths. Color two sixths.

Which third of a triangle that you colored is largest? ______________________

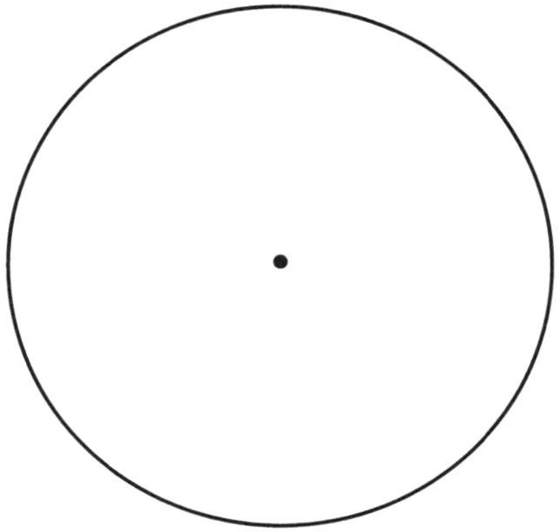

Divide the circle into thirds. Color one third.

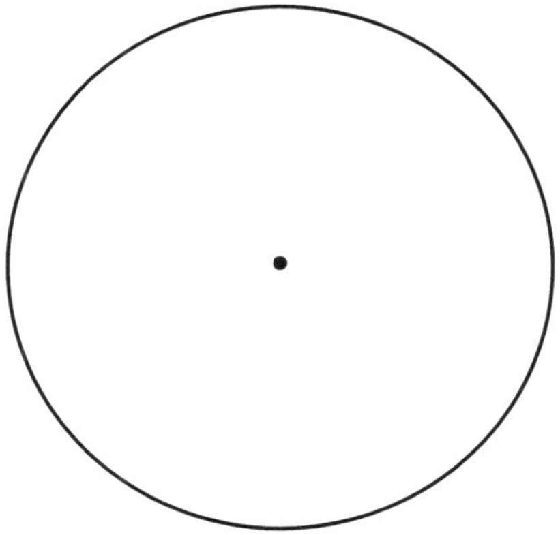

Divide the circle into sixths. Color two sixths.

Which is more, $\frac{1}{3}$ or $\frac{2}{6}$? __________

Name:________________________ Date: ________________________

Draw the divided equilateral triangles on the lines below.

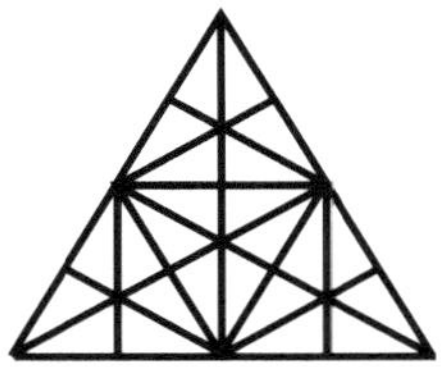

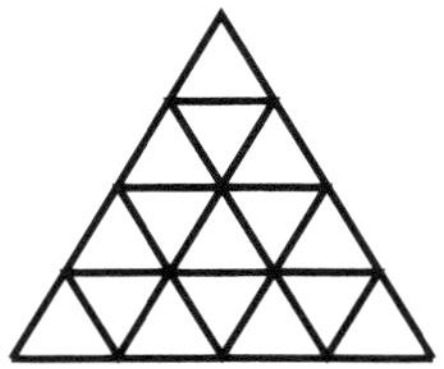

How many small congruent triangles are in the triangle above? ______

How many small congruent triangles are in the triangle at the right? _____

Name: ______________ Date: ______________

Draw a star according to the instructions below.

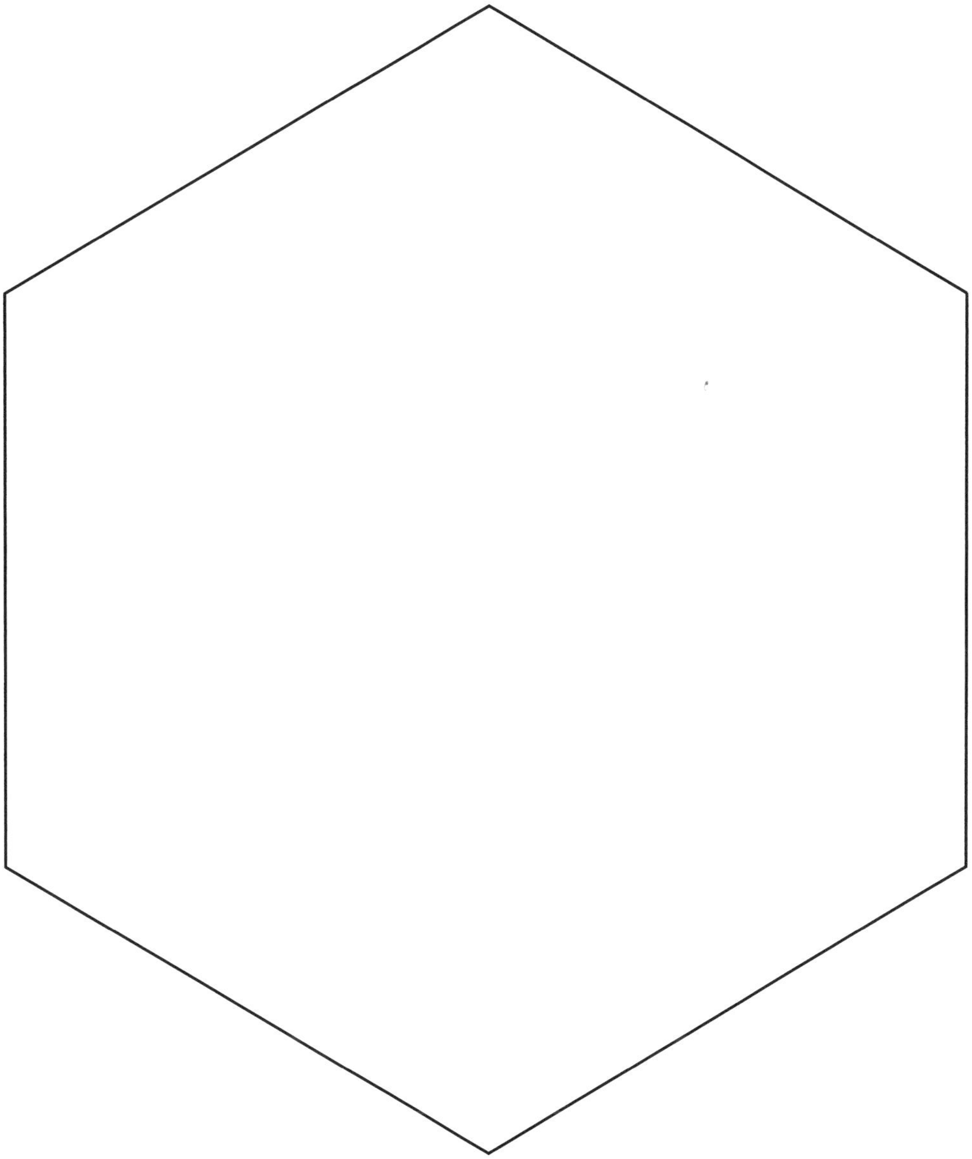

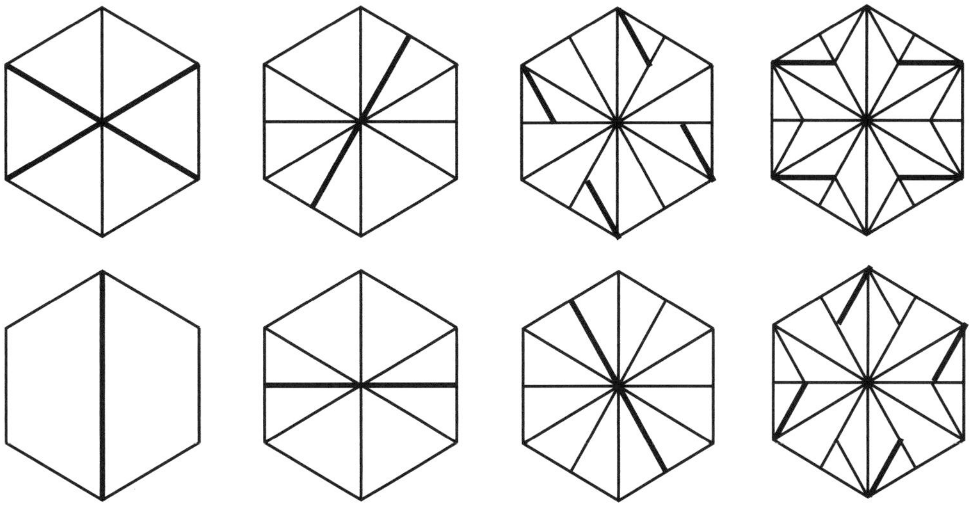

Name: ______________ Date: ______________

Draw a star according to the instructions below.

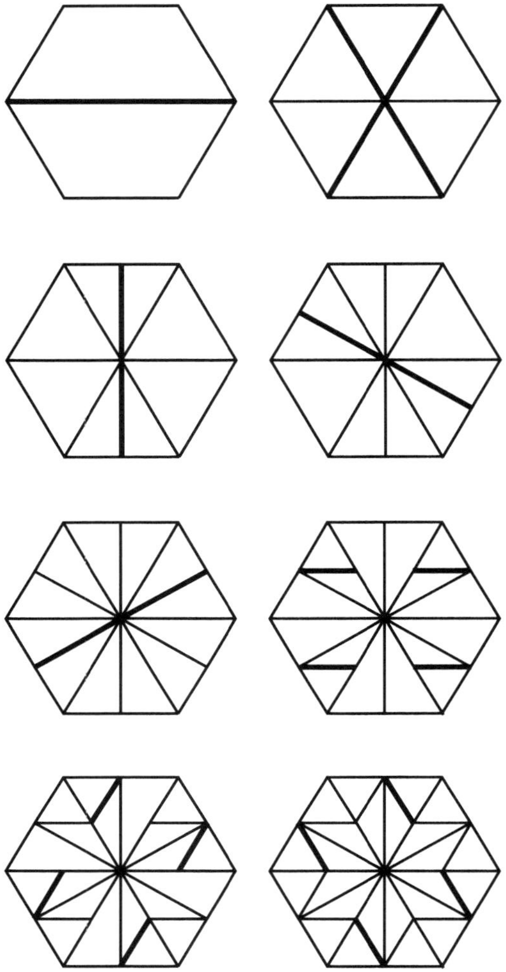

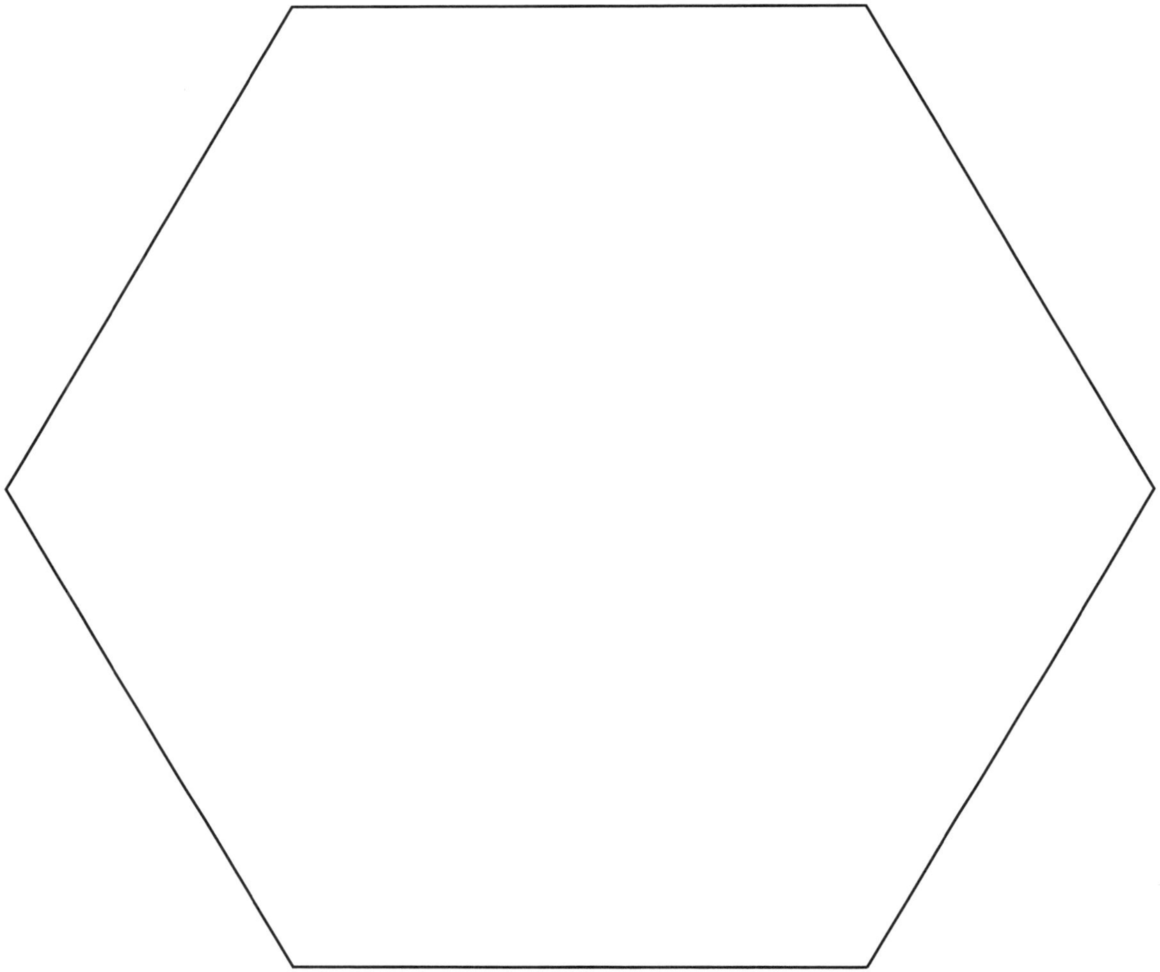

Find the perimeter of the hexagon in inches. ______________

Find the perimeter of the star in inches. ______________

Name: ____________________ Date: ____________________

1. Outline 3 equilateral triangles of different sizes.

2. Outline a triangle that is not equilateral.

3. Outline a rhombus.

4. Outline a parallelogram that is not a rhombus.

5. Outline 2 trapezoids of different sizes.

6. Outline 3 regular hexagons of different sizes.

7. Outline 1 hexagon that is not regular.

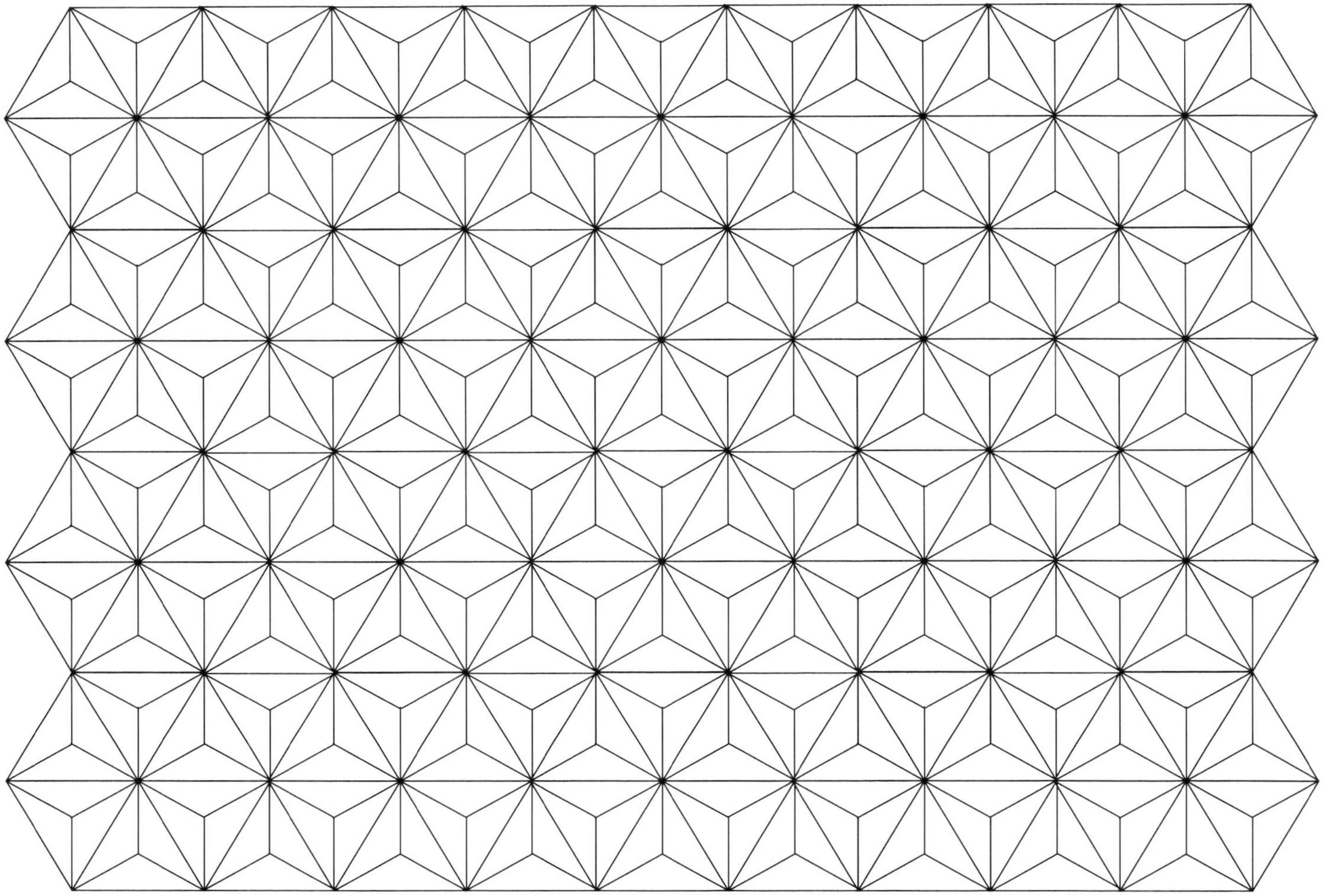

Name: ______________________ Date: ______________________

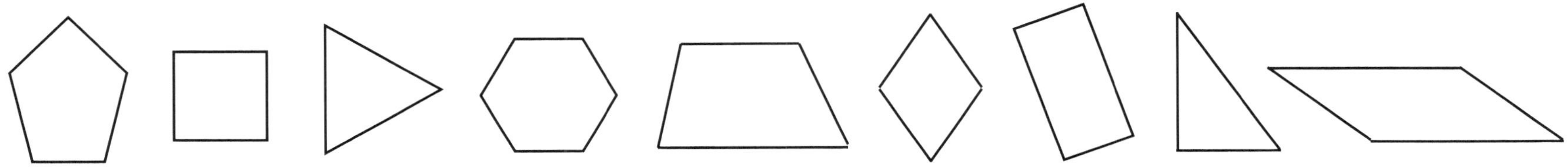

A B C D E F G H I

Name of polygon	Which letter?	Are all sides equal?	Are all angles equal?	Regular polygon?	Does it have symmetry?	Number of sets of parallel lines
Rectangle						
Rhombus						
Regular hexagon						
Pentagon						
Right triangle						
Equilateral triangle						
Parallelogram						
Square						
Trapezoid						

Use your chart to help you answer some of the following questions.

How many of the polygons have all sides equal? ____

How many of the polygons have all angles equal? ____

How many of the polygons are regular? ____

Which polygon has the most sets of parallel lines? ____

How many of the polygons are quadrilaterals? ____

How many of the polygons are not quadrilaterals? ____

Name: ______________________

Date: ______________________

Write only the answers.

____

____

____

Write the answers.

$4\frac{1}{2} + 2\frac{1}{2} =$ ______

$5 \times 4 =$ ______

$1 - \frac{1}{2} =$ ______

Draw lines to match each drawing to the words.

rhombus

hexagon

equilateral triangle

right triangle

cm

pentagon

inch

trapezoid

Sam had 5729 trees and planted 367 more trees. How many trees does Sam have now? ______________

Divide the rectangle into equal quarters three different ways.

Name: ____________________ Date: ____________________

Color your own designs.

Name: ______________________________

Date: ______________________________

Write only the answers.

____

____

____

Write the answers.

$3\frac{1}{2} + 4\frac{1}{2} =$ ______

$4 \times 3 =$ ______

$1\frac{1}{2} - \frac{1}{2} =$ ______

Draw lines to match each drawing to the words.

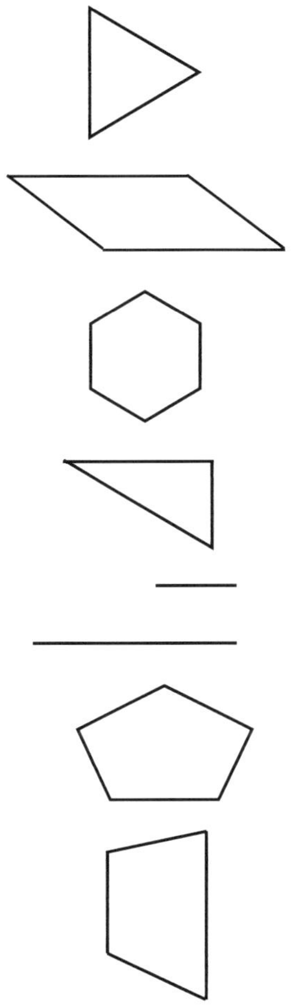

parallelogram

hexagon

equilateral triangle

30-60 triangle

centimeter

pentagon

inch

trapezoid

The library received 256 new books. They already had 7692. How many books does the library have now? ______________

Divide the square into equal fourths three different ways.

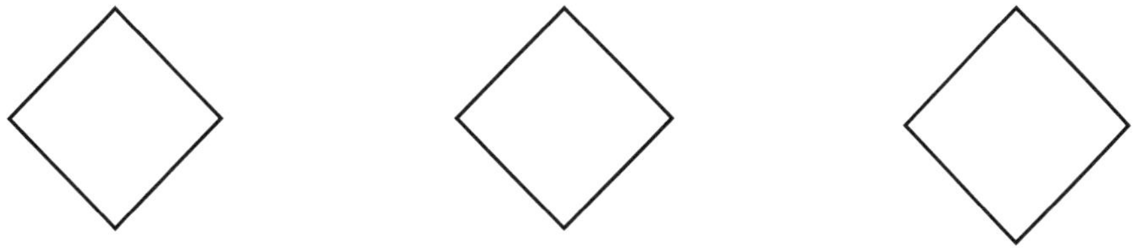

Name: ______________________ Date: ______________________

Draw the arrow for 58.

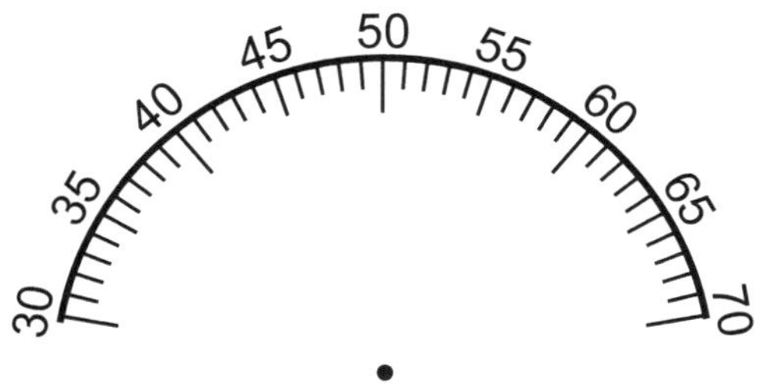

Draw the arrow for 37.

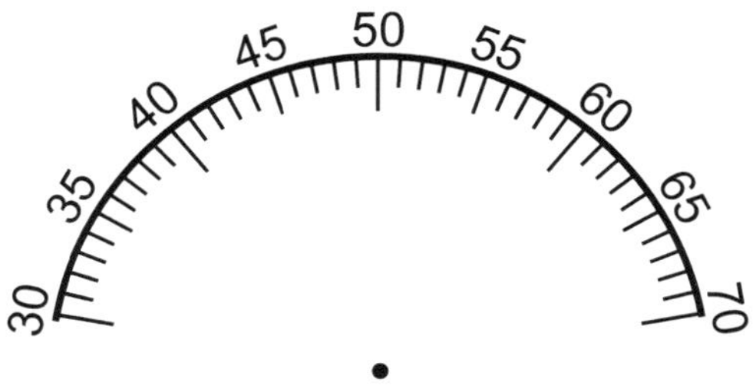

Read the dial. _____

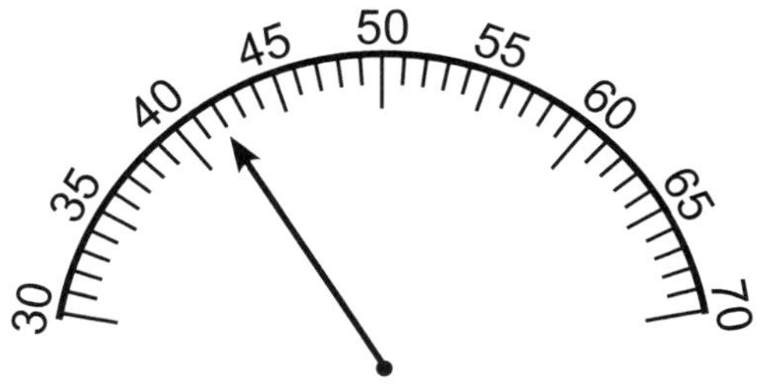

Draw line for 18.

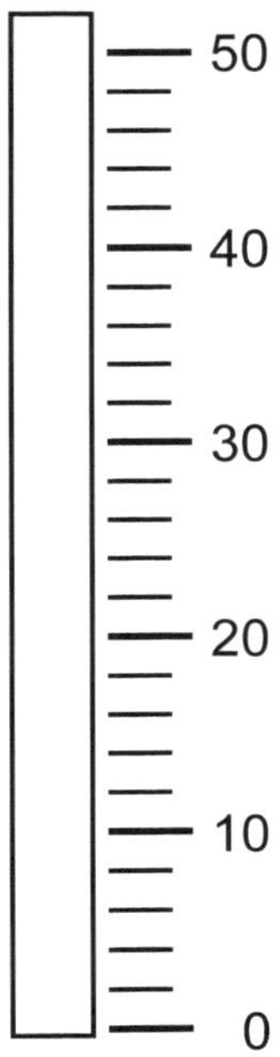

Draw line for 76.

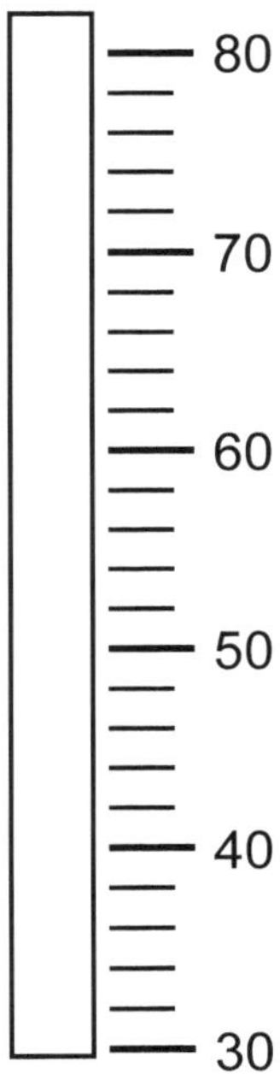

Read. _____

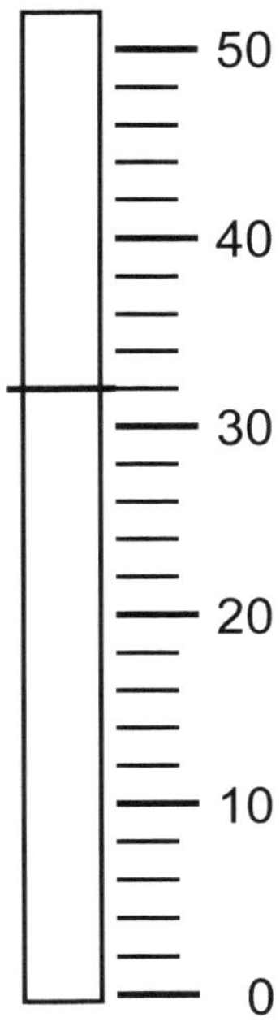

Read. _____

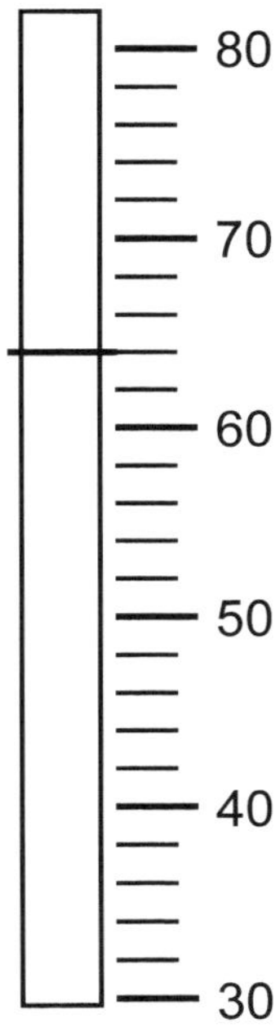

Draw line at 45.

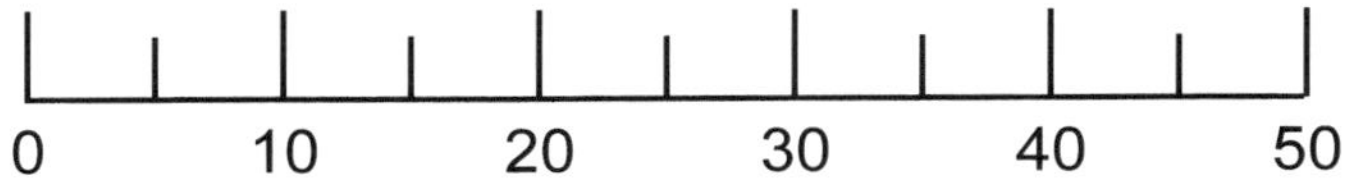

Draw line at 35.

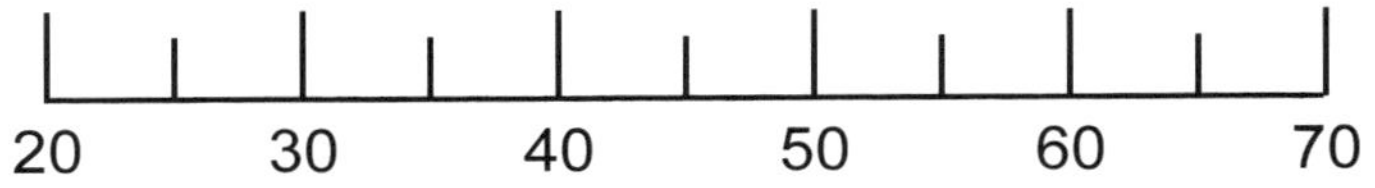

Read. _____

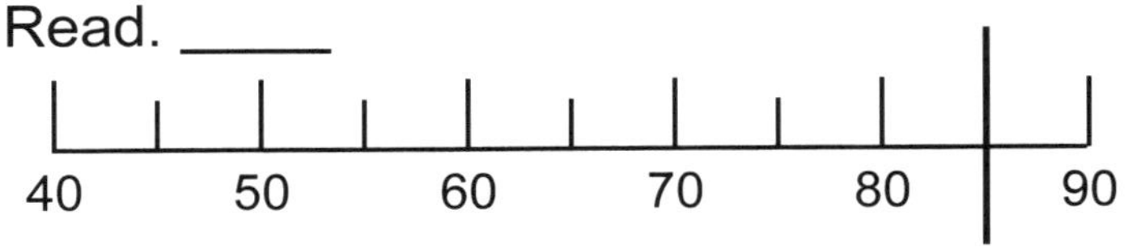

Read. _____

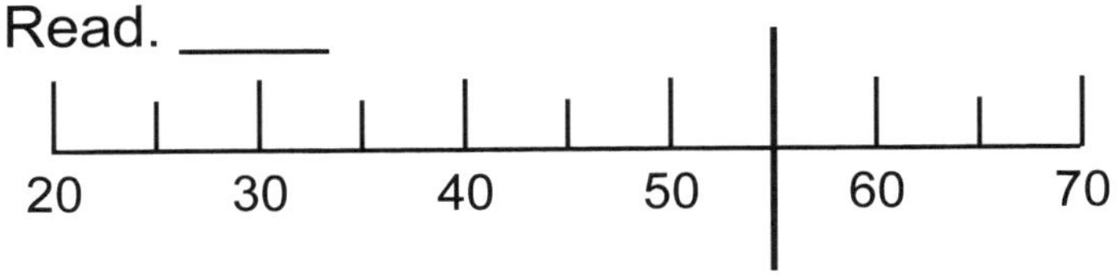

Name: ______________________________

Date: ______________________________

Use drawing tools to recreate the clock. Draw the hour marks and write the hour numbers.

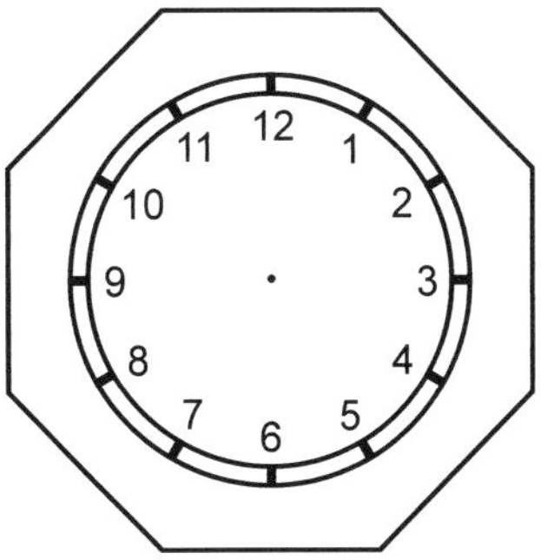

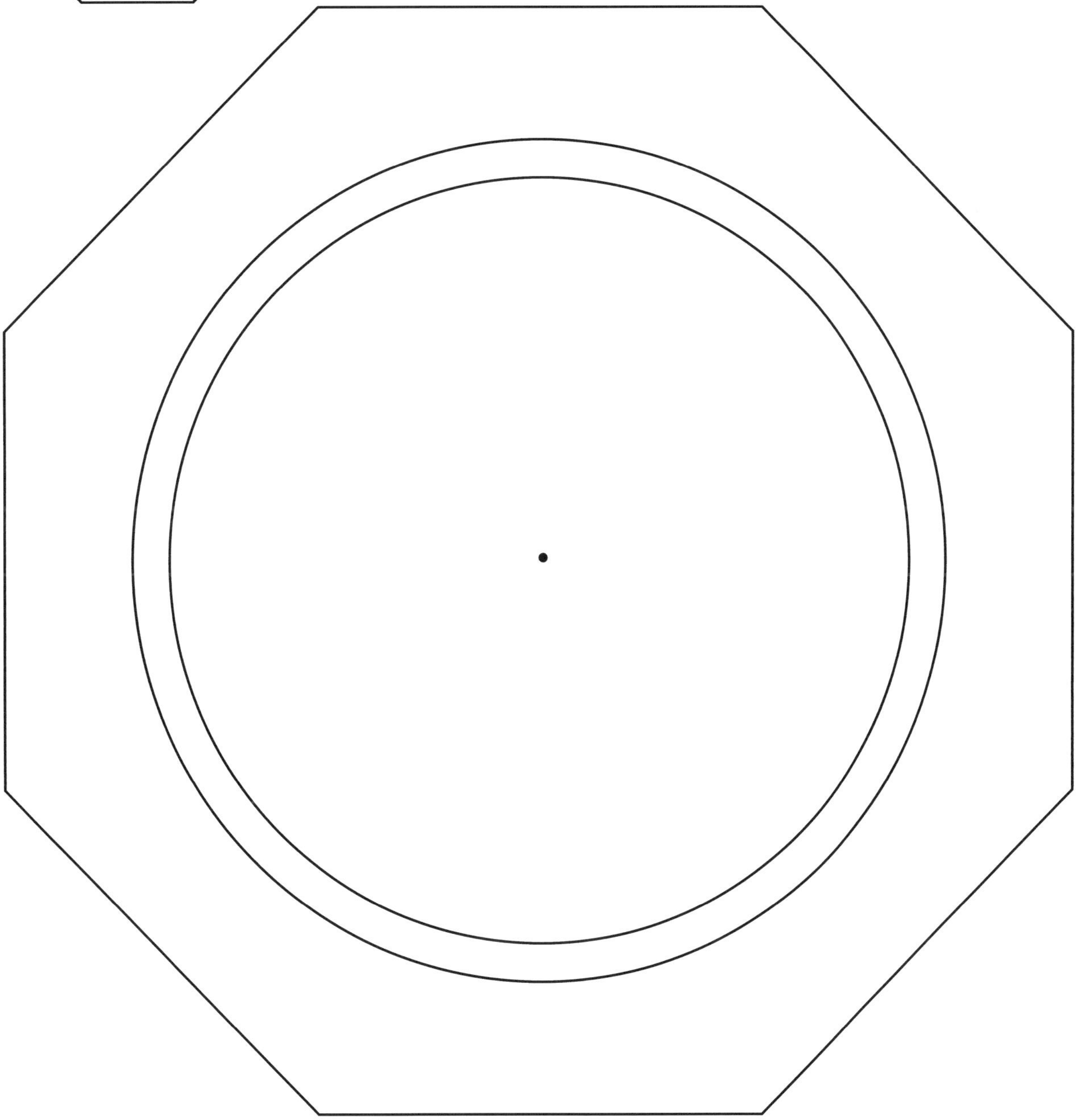

Name: ____________________ Date: ____________________

Draw lines to connect the activities to the hour of the day. Write your own activities on the blank lines.

Hours of the Day

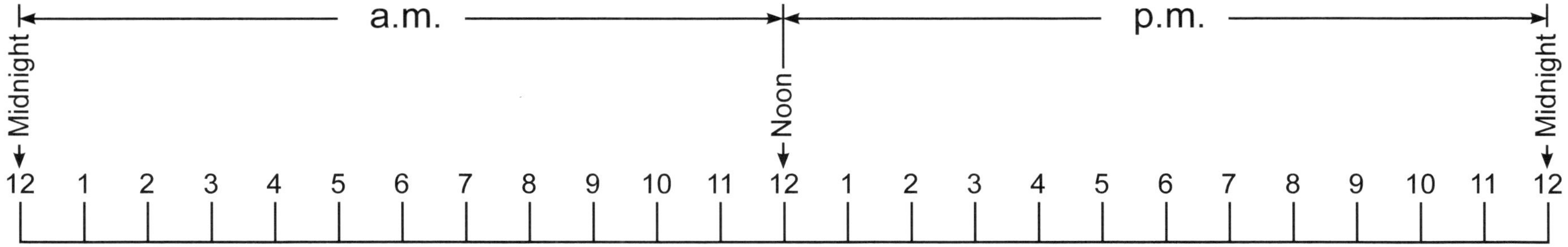

Eat breakfast ____ Learn math ____ Get up ____ Eat lunch ____

Eat dinner ____ Play outside ____ Go to bed ____

____________________ ____________________

1. When does a day start? ____________________
2. What is special about noon? ____________________
3. How many hours in a day? ____________________
4. When do we use a.m.? ____________________
5. When do we use p.m.? ____________________
6. How long does it take the hour hand to go from 2 to 3? __________

Name: ______________________________

Date: ______________________________

Draw lines to match the digital clocks to the analog clocks.

8:00

8:05

8:10

8:15

8:20

8:25

8:30

8:35

8:40

8:45

8:50

8:55

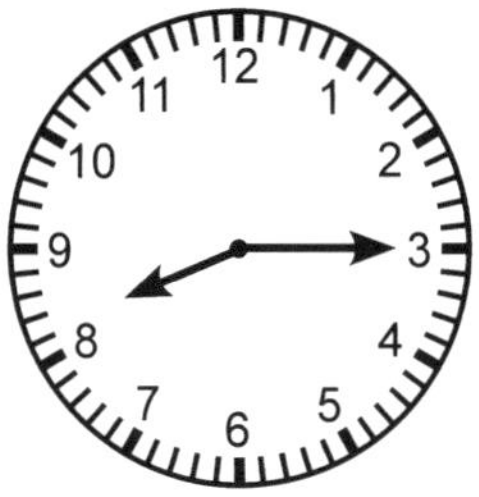

Name: ______________________________

Date: ______________________________

Write the time.

A.

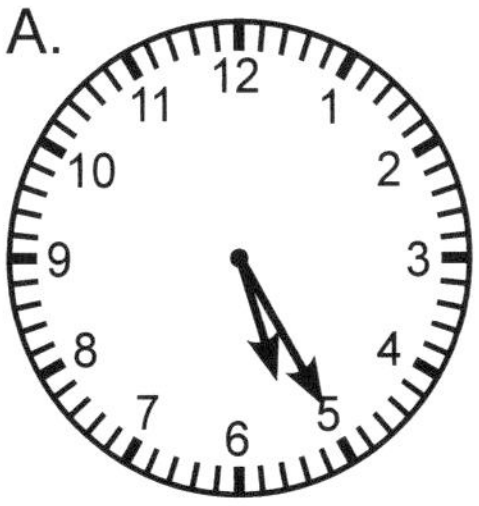

______

B.

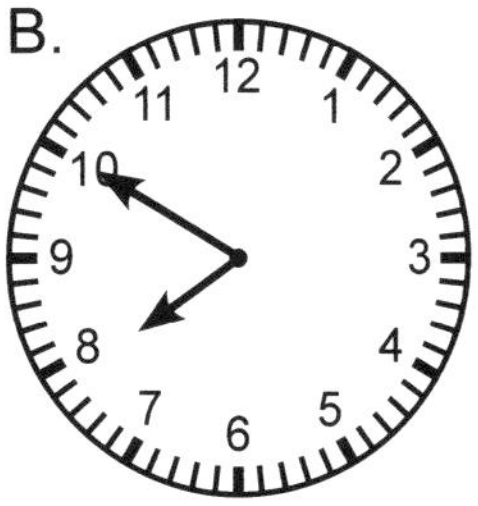

______

C.

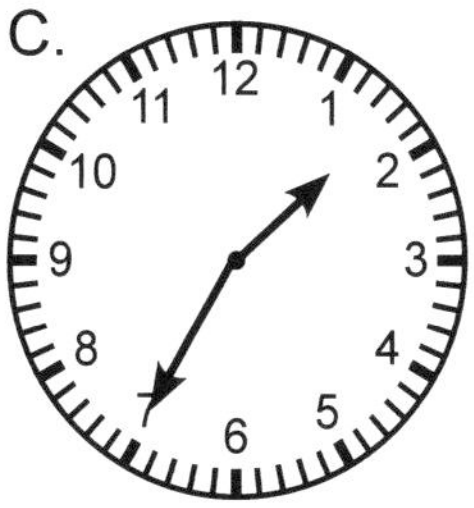

______

D.

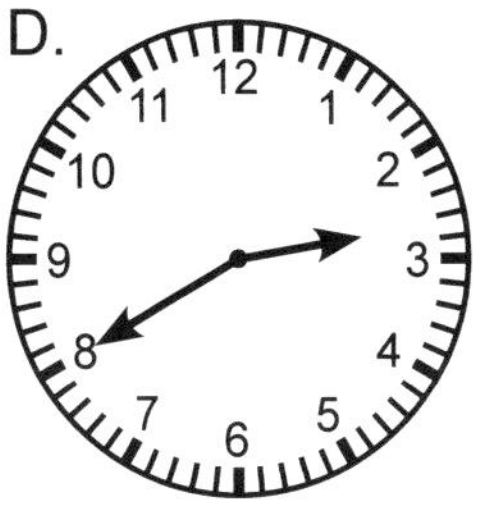

______

E.

______

F.

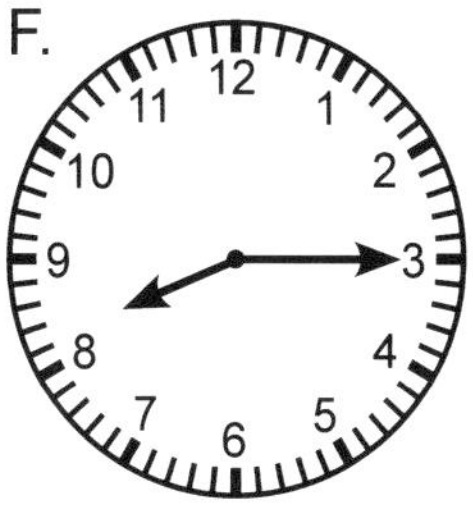

______

G.

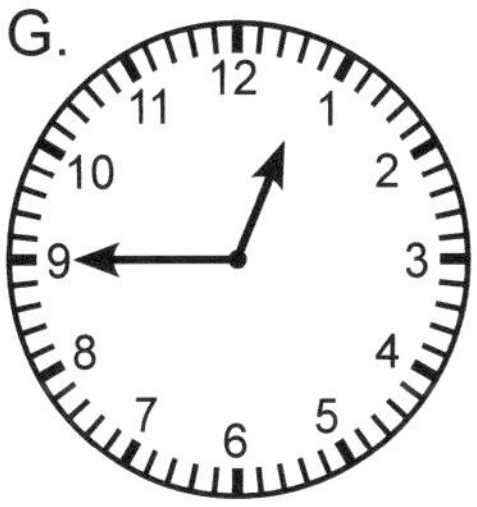

______

H.

______

I.

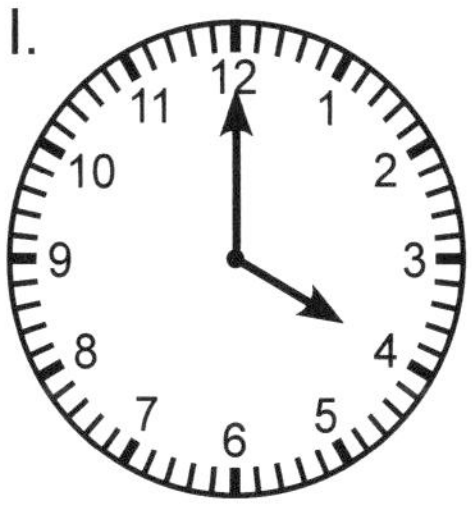

______

Draw the hands.

J.

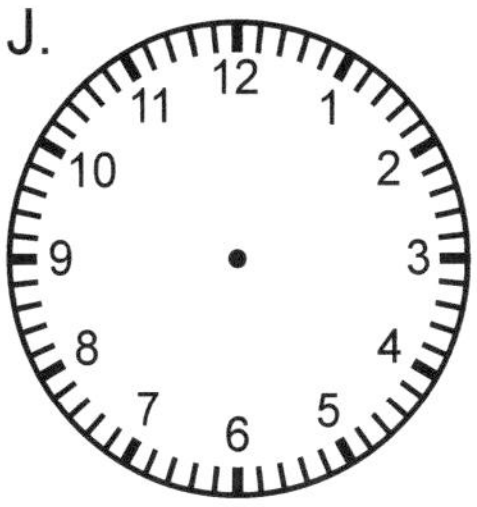

3:30

K.

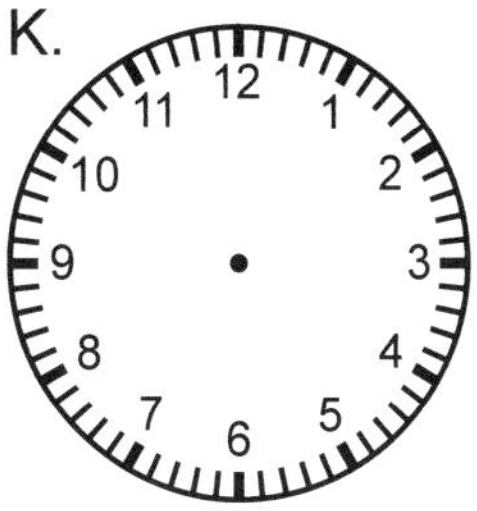

9:55

L.

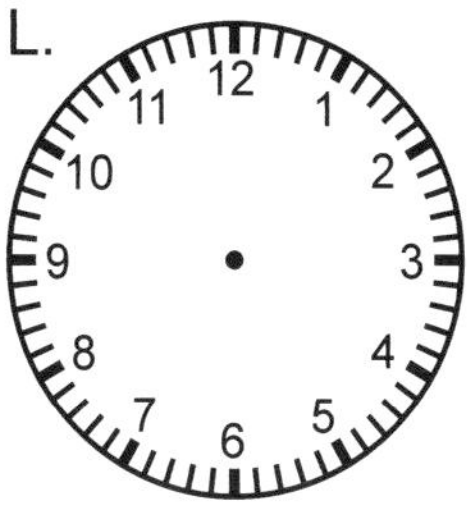

11:10

Which clock shows a quarter after? _____

Which clock shows half past? _____

Which clock shows a quarter to? _____

Name: ______________________________

Date: ______________________________

Draw lines to match the digital clocks to the analog clocks.

10:02

10:07

10:14

10:16

10:22

10:27

10:31

10:39

10:43

10:48

10:54

10:57

Name: ______________________________

Date: ______________________________

Write only the answers.

____

____

____

Write the answers.

$\frac{1}{2}$ hour + $\frac{1}{2}$ hour = ______

13 – 4 = ______

11 – 6 = ______

Write the time.

__________ __________ __________

Draw the hands.

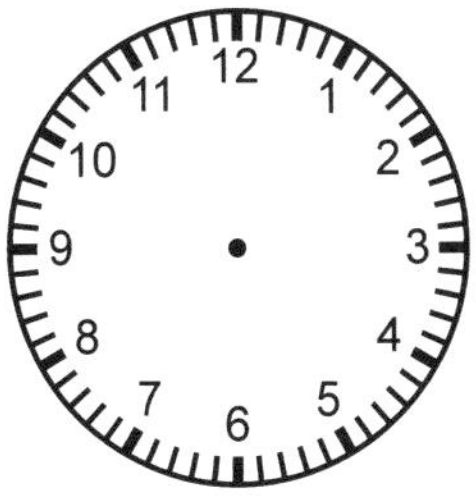

7:10

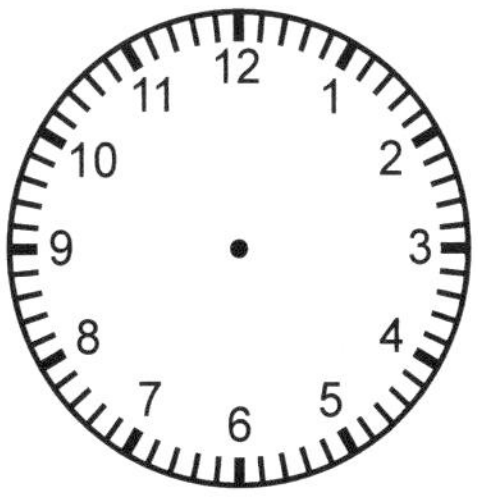

3:22

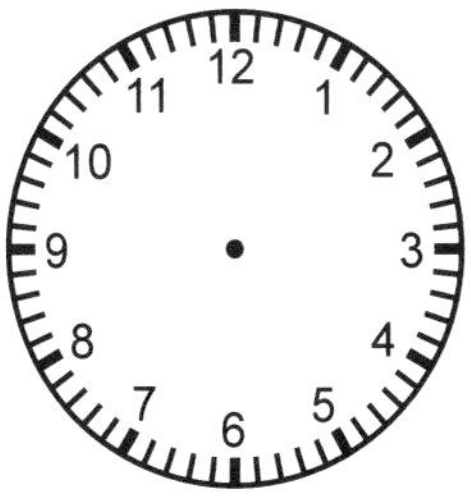

5:48

Solve the problem.

Alex jumped rope 499 times and Adrian jumped 501 times. Who jumped more and how much more? ______________________________

Use > for “is after.” Use < for “is before.” Use = for “is the same as."

half past 10 _____ 10:15

11:45 a.m. _____ 1:10 p.m.

quarter to 1 _____ 12:45

quarter to 2 _____ quarter after 2

Write this number in words.

6411 ______________________________

Name: ______________________________

Date: ______________________________

Write only the answers.

____

____

____

Write the answers.

1 hour – $\frac{1}{2}$ hour = ______

14 – 5 = ______

12 – 7 = ______

Write the time.

__________ __________ __________

Draw the hands.

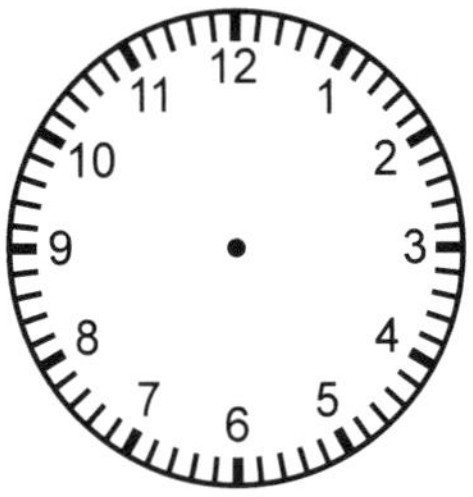

3:45

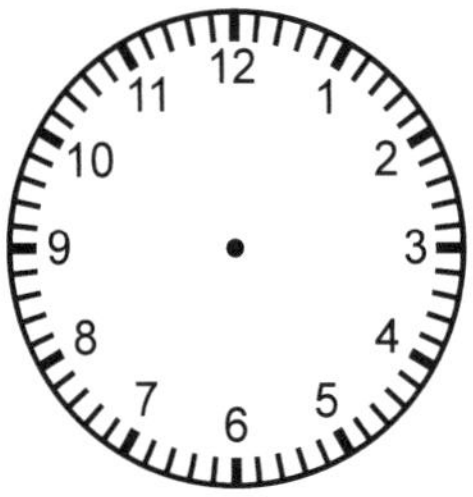

1:29

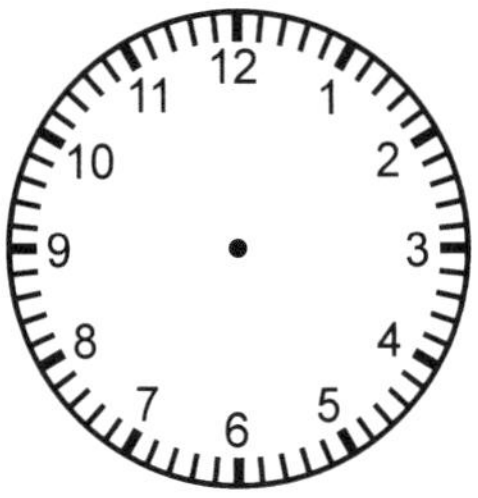

9:13

Solve the problem.

Tyler read 400 pages altogether. Val read 390 pages. Who read more pages and how many more? ______________________________

Use > for "is after." Use < for "is before." Use = for "is the same as."

quarter to 7 _____ 6:45

3:45 p.m. _____ 6:50 a.m.

quarter after 9 _____ 9:45

half past 3 _____ 3:30

Write this number in words.

4213 ______________________________

Name: ____________________________

Date: ____________________________

Write the equations and solve the problems.

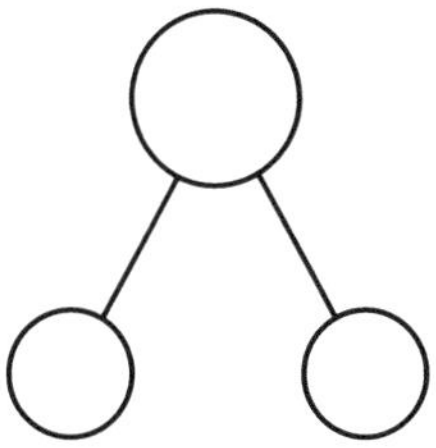

1. Mr. Black is 6 feet tall. His son is 4 feet tall. How much taller is the father?

____

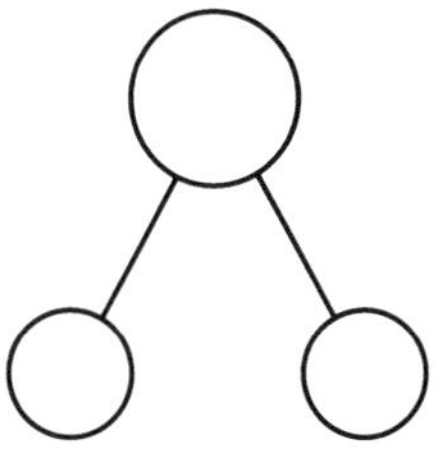

2. Mrs. Jackson is 170 cm tall. Her daughter is 119 cm tall. How much taller is the mother?

____

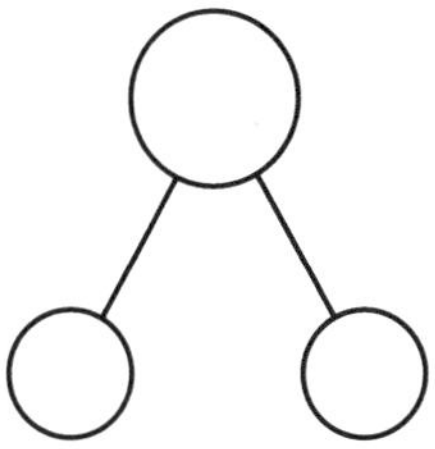

3. Jasmine has five pillows. Oliver has four more pillows than Jasmine. How many pillows does Oliver have?

____

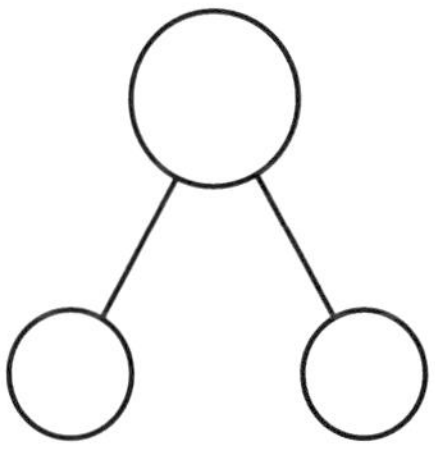

4. Logan has 12 more cherries than Matt. Matt has 25 cherries. How many cherries does Logan have?

____

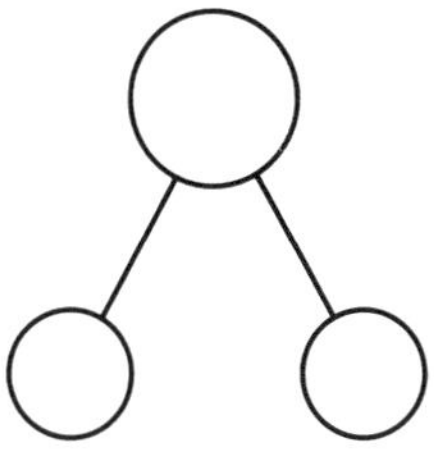

5. Shauna has 3 more flowers than Jacob. Shauna has 5 flowers. How many flowers does Jacob have?

____

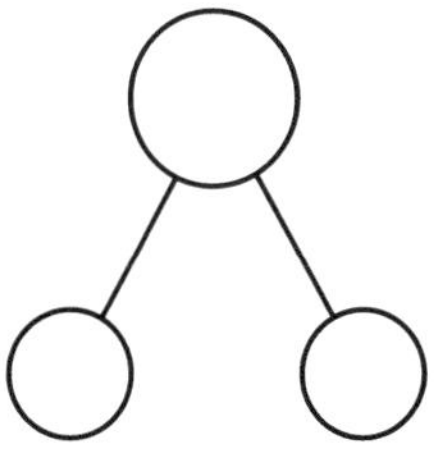

6. James has 20 grapes. James has 11 more grapes than Lily. How many grapes does Lily have?

____

Name: ____________________

Date: ____________________

Write the equations and solve the problems.

1. Brandon is 2 cm shorter than his twin, Kayla. Brandon is 149 cm tall. How tall is Kayla?

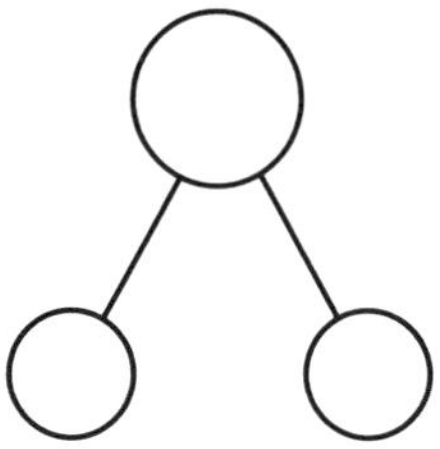

____

2. Michaela practiced her violin 21 hours last month and 18 hours this month. How much less time did she practice this month?

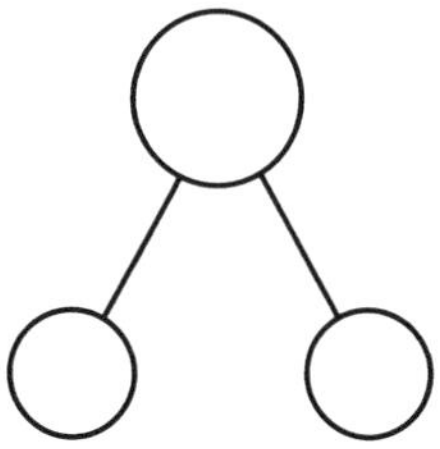

____

3. Joshua's team scored 9 fewer points than Noah's team. Joshua's team scored 43 points. How many points did Noah's team score?

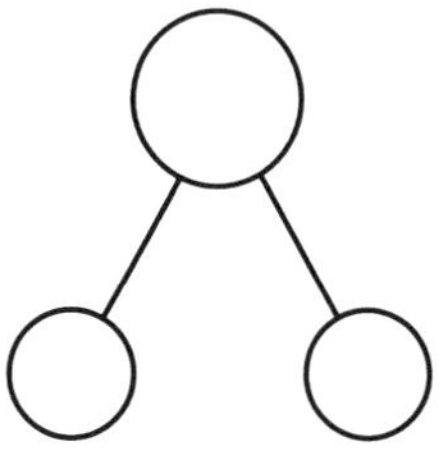

____

4. Joshua's team scored 9 fewer points than Noah's team. Noah's team scored 43 points. How many points did Joshua's team score?

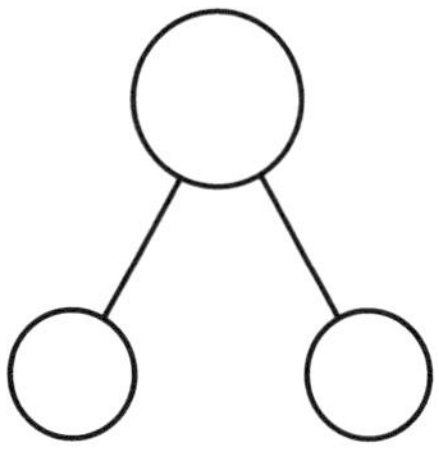

____

5. Emily's apple weighs 80 grams. Sophie's apple weighs 6 grams less than Emily's apple. How much does Sophie's apple weigh?

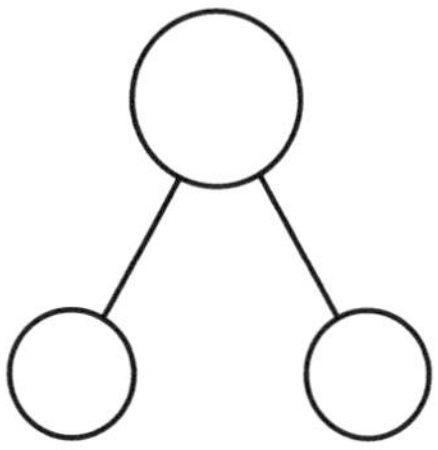

____

6. The Wang family traveled 18 miles to a game. The Soto family traveled 27 miles. How many miles less did the Wang family travel?

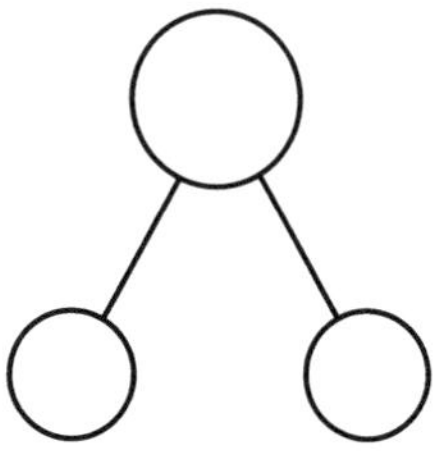

____

Name: ______________________________

Date: ______________________________

Write the equations and find the answers with the base-10 cards.

1. The city of Logan, Utah is 4534 feet above sea level. The city of Billings, Montana is at 3123 ft. How much higher is Logan than Billings?

____

2. The Appalachian Mountains have a peak of 6684 feet. Sugarloaf Mountain in Maryland is 1273 feet high. How much taller is the Appalachian peak?

____

3. The distance between New York and Hawaii is 4858 miles. The distance between New York and England is 3296 miles. How much closer is New York to England than to Hawaii?

____

Subtract the following with base-10 cards and write the differences.

4.

	4	5	3	4
–	2	4	1	8

5.

	5	0	7	2
–	2	5	4	5

6.

	6	4	9	1
–	5	7	8	5

7.

	7	0	8	0
–	3	8	2	9

8.

	8	2	1	5
–	5	4	3	7

9.

	9	2	4	7
–	5	6	8	5

Name: ______________________________

Date: ______________________________

Subtract using side 2 of the abacus.

	4	9	9	5
–		1	1	1
–		2	2	2
–		3	3	3
–		4	4	4
–		5	5	5
–		6	6	6
–		7	7	7
–		8	8	8
–		9	9	9

	4	9	9	5
–		9	9	9
–		8	8	8
–		7	7	7
–		6	6	6
–		5	5	5
–		4	4	4
–		3	3	3
–		2	2	2
–		1	1	1

If you would like, make up your own numbers to subtract.

Name: ______________________________

Date: ______________________________

Subtract using your abacus.

1.

	6	8	2	9
–	2	6	3	7

2.

	7	0	9	4
–	3	5	2	8

3.

	6	8	3	4
–	1	5	6	3

4.

	3	4	5	8
–	1	8	2	5

5.

	4	0	5	8
–	2	7	4	9

6.

	9	2	4	7
–	5	6	6	5

7.

	4	5	3	4
–	2	4	1	8

8.

	5	0	7	2
–	2	5	4	5

9.

	6	4	9	1
–	5	7	8	5

10.

	7	0	8	0
–	3	8	2	9

11.

	8	2	1	5
–	5	4	3	7

12.

	9	2	4	7
–	5	6	8	5

Name: ______________________________

Date: ______________________________

Subtract.

	3	5	7	0
–		3	5	7
–		3	5	7
–		3	5	7
–		3	5	7
–		3	5	7
–		3	5	7
–		3	5	7
–		3	5	7
–		3	5	7
–		3	5	7

Write a 4-digit number with a 0 in the ones place. Subtract the number without the 0.

Solve the problem.

The total distance from Windsor, Canada, to Toronto is 370 km. The distance from Windsor to London is 183 km. The distance from London to Hamilton is 119 km. What is the distance from Hamilton to Toronto?

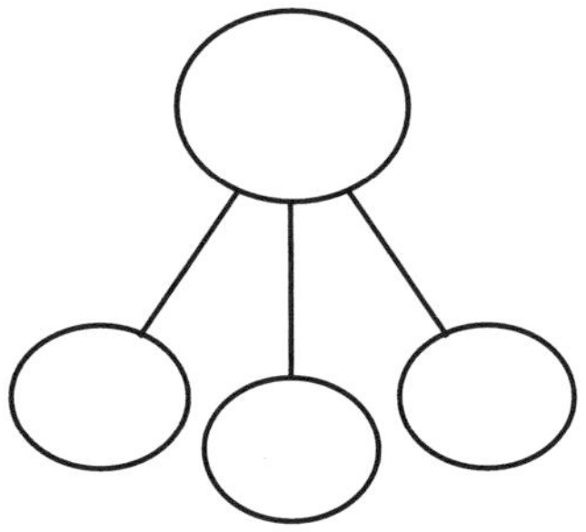

____

Name: ______________________________

Date: ______________________________

Write only the answers.

____

____

____

Write the answers.

38 + 78 = ______

206 + 49 = ______

306 – 293 = ______

Write the time.

____

Draw the hands.

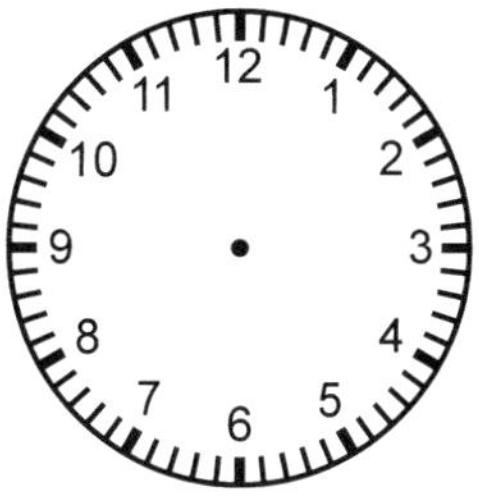

12:20

Subtract.

	4	0	8	5
–	3	8	6	9

Solve the problem.

The Scenic bus drove 2847 km. The Touring bus drove 529 km farther than the Scenic bus. How far did the Touring bus drive?

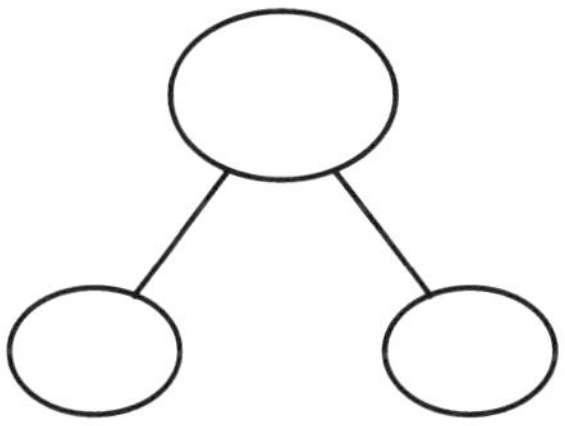

Write >, <, or = on the lines.

2086 _____ 2806

5371 – 100 _____ 5271

4253 – 256 _____ 4253 + 256

399 + 2 _____ 301

54 – 11 _____ 54 – 13

6000 – 1 _____ 5999

Explain what is wrong with this thinking.

64 – 27 = 43. 60 – 20 is 40 and 7 – 4 is 3. So the answer is 40 + 3 = 43.

____

____

Name: ______________________________

Date: ______________________________

Write only the answers.

____

____

____

Write the answers.

47 + 86 = ______

307 + 58 = ______

504 – 497 = ______

Write the time.

____

Draw the hands.

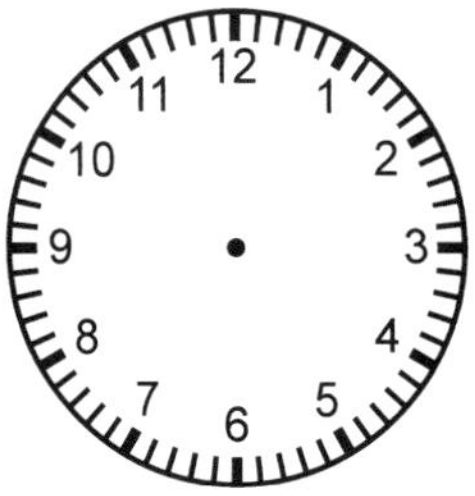

9:15

Subtract.

	7	8	8	6
–	5	8	6	7

Solve the problem.

The Rail train traveled 1847 km. The Trax train traveled 2073 km. How much farther did the Trax train travel?

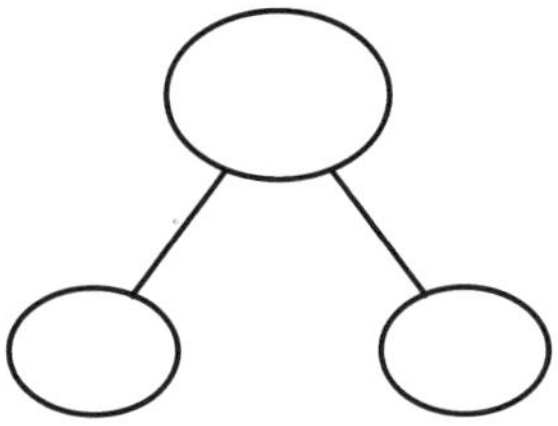

Write >, <, or = on the lines.

5901 _____ 5091

7126 – 1000 _____ 5126

8146 – 789 _____ 8146 + 789

499 + 2 _____ 501

47 – 14 _____ 47 – 12

7000 – 1 _____ 7999

Explain what is wrong with this answer.

47 – 29 = 22. 40 – 20 is 20 and 9 – 7 is 2. So the answer is 20 + 2 = 22.

____

____

Name: ____________________ Date: ____________________

Write the total value of the coins.

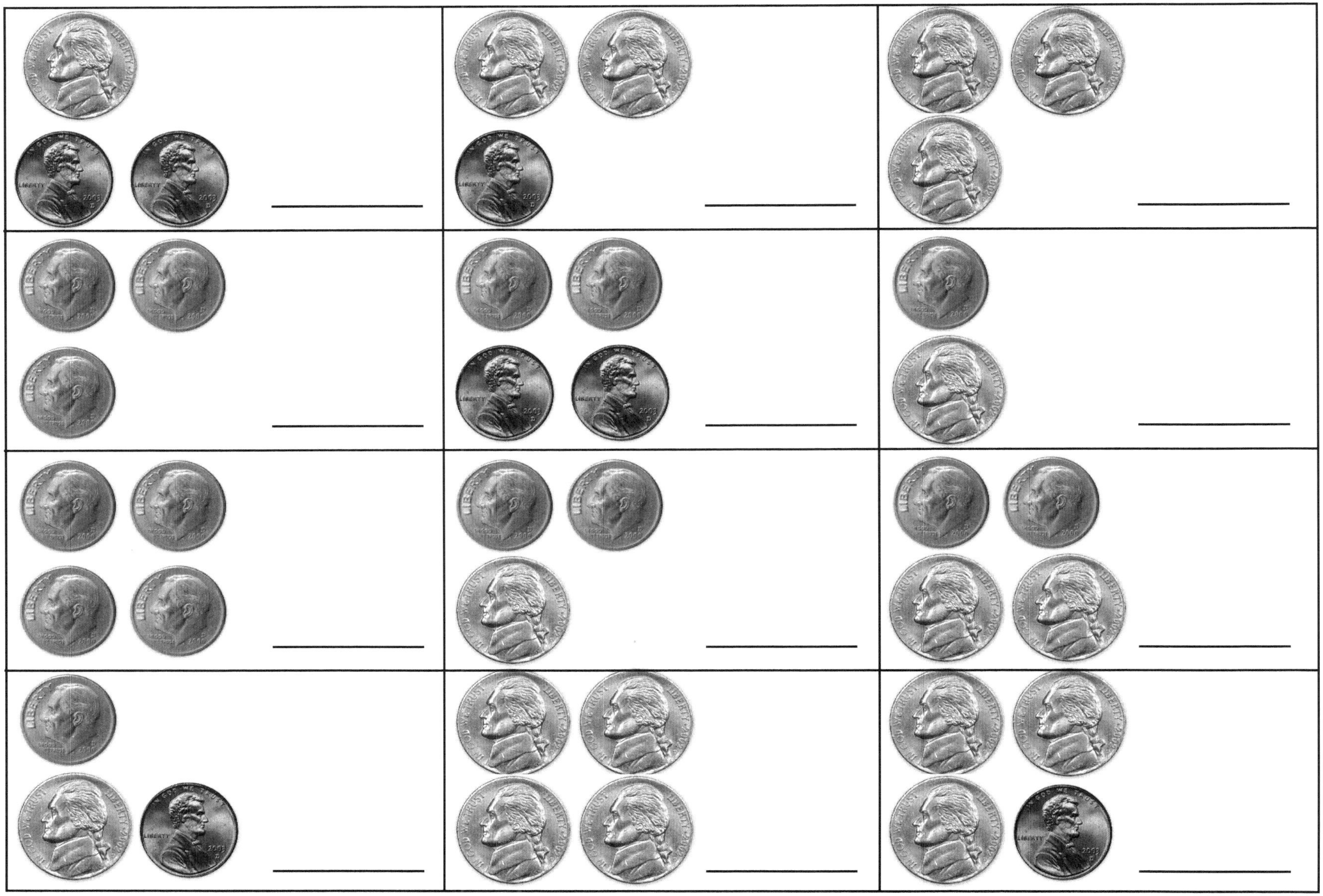

Name: ______________________________

Date: ______________________________

Choose three coins to make the amount given in each row.

	Half-dollars	Quarters	Dimes	Nickels	Pennies
7¢				1	2
11¢					
16¢					
21¢					
15¢					
75¢					
76¢					
31¢					
25¢					
27¢					
56¢					
36¢					
61¢					
45¢					

Name:______________________ Date: ______________________

List the coins needed to make a dollar using the number of coins given at the left.

Number of coins	Half-dollars	Quarters	Dimes	Nickels
2				
3				
4				
5				
6				
7				
8				
9				
10				

Number of coins	Half-dollars	Quarters	Dimes	Nickels
11				
12				
13				
14				
15				
16				
17				
18				
19				
20				

Name: ______________________________

Date: ______________________________

Write the change needed to buy what is listed. Items are bought with 50¢.

A toy costing 38¢: 39, 40, 50¢

A balloon costing 21¢: ______________________________

A banana costing 15¢: ______________________________

A book costing 29¢: ______________________________

Two apples costing 18¢ each: ______________________________

Three pencils costing 11¢ each: ______________________________

Write the change needed to buy what is listed. Items are bought with 1 dollar.

A pen costing 73¢: ______________________________

An orange costing 34¢: ______________________________

A token costing 85¢: ______________________________

A notebook costing 59¢: ______________________________

Four apples costing 20¢ each: ______________________________

Two toys costing 22¢ each: ______________________________

Make up some of your own purchases.

____

____

____

Name: ______________________ Date: ______________________

Write the value of the coins using a dollar sign and decimal point.

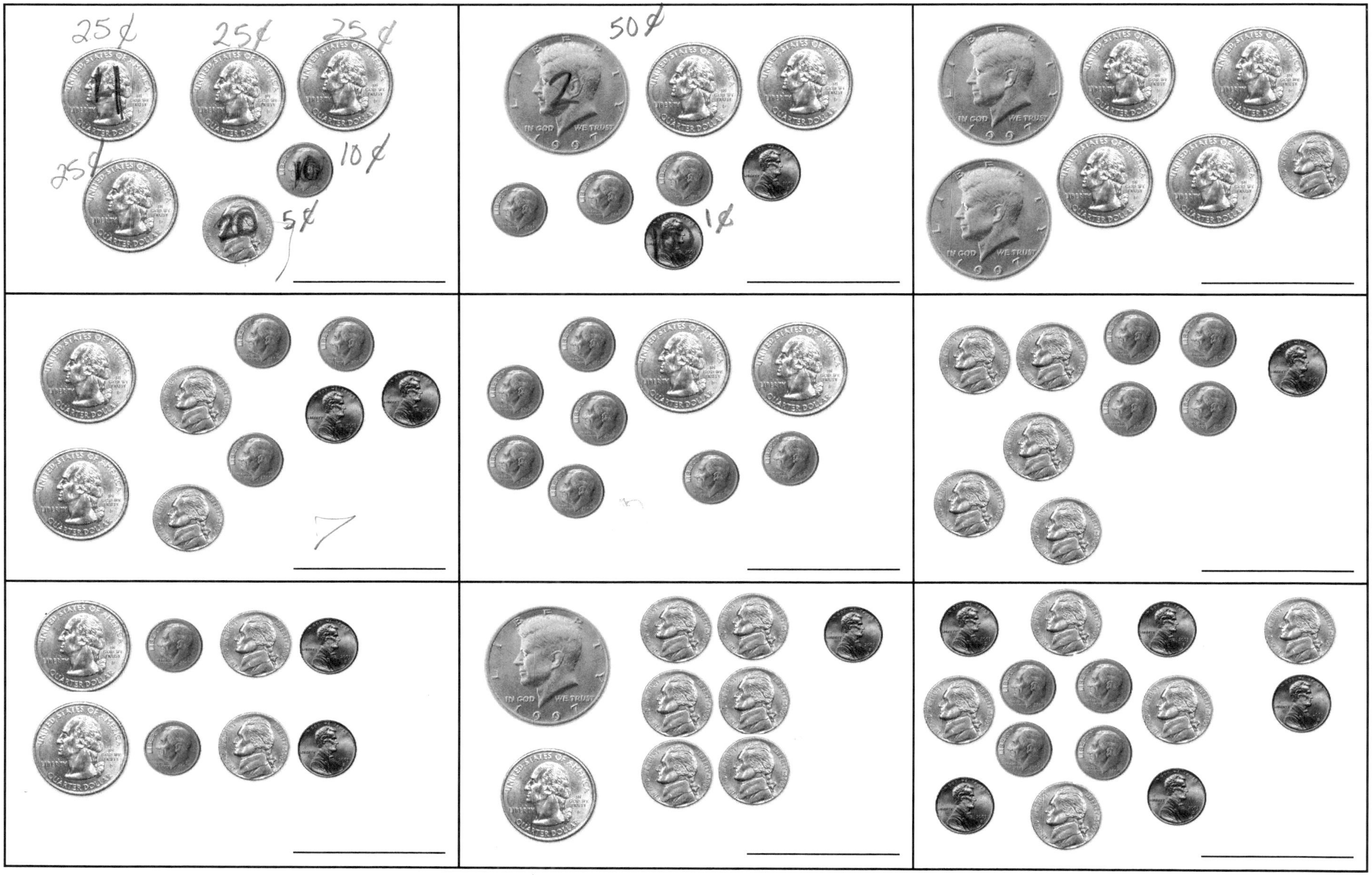

Name: Jaxon

Date: 11/19/24

Solve the problems using the part-whole circle sets. Write the equations.

1. Amy has two quarters, three dimes, and two nickels. How much money does she have? Is it more or less than a dollar?

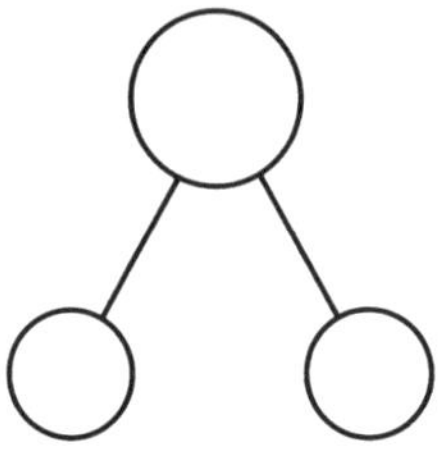

____

2. Todd wants to buy a gift that costs one dollar. He has 3 quarters and 4 nickels. How much more money does he need? What coins could he use?

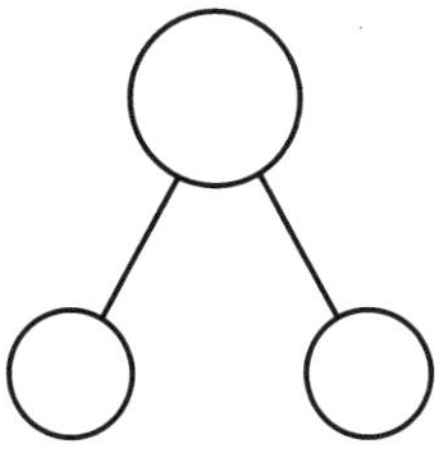

____

3. Belle is buying two bananas. Each banana cost 39¢. What is her change from a dollar?

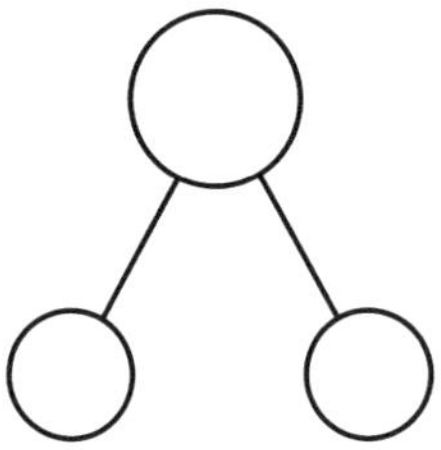

____

4. Logan has 56¢. He has one quarter, five nickels, and some pennies. How many pennies does he have?

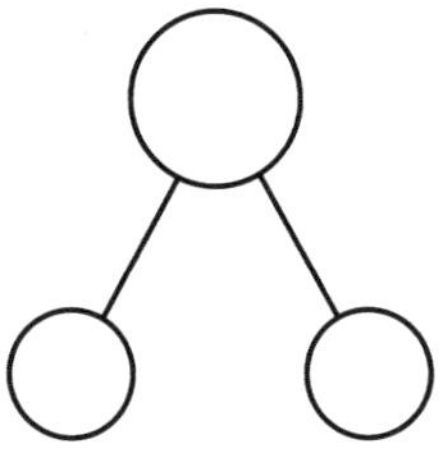

____

5-8. Add. Use the dollar sign, $, for your answers.

56¢ + 13¢ + 17¢ + 20¢ = _______ $2 + 49¢ + 10¢ + 5¢ = _______

79¢ + 29¢ + $1.01 + 1¢ = _______ 25¢ + 25¢ + 50¢ + 50¢ = _______

Name: ____________________________________

Date: ____________________________________

Write only the answers.

____

____

____

Write the answers.

3 dimes + 19¢ = ______

$3.39 + $0.21 = _______

$1.31 – 32¢ = _______

Add 284 + 1836.

Add.

	4	1	8	5
+	3	8	6	9

Subtract.

	7	7	4	1
–	5	7	6	9

Solve the problem.

Lacey received $1.09 on Wednesday and three quarters and three dimes on Thursday. How much more money did Lacey receive on Wednesday? Write your answer two ways using a ¢ and $.

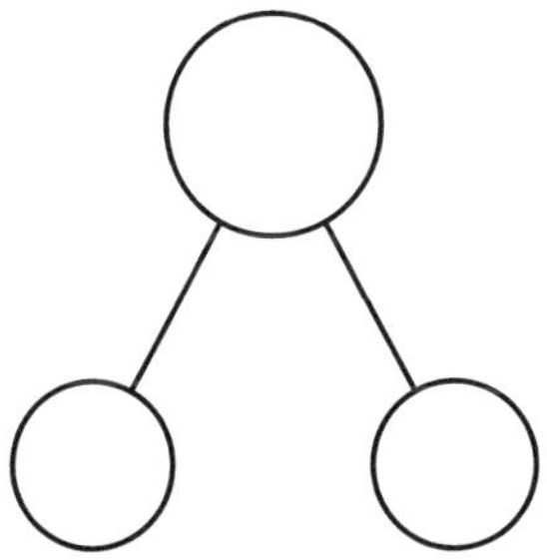

____

Write >, <, or = on the lines.

1 nickel _____ 1 dime

1 dollar _____ 4 quarters

$0.76 _____ 76¢

6 dimes _____ 11 nickels

3 quarters _____ 8 dimes

Complete the equations.

4 nickels = _____ dimes

_____ pennies = 3 quarters

$1.45 = $1 + _____ cents

3 nickels + _____ dimes = 1 quarter

4 dimes + _____ nickels = 2 quarters

Name: ______________________________

Date: ______________________________

Write only the answers.

____

____

____

Write the answers.

2 dimes + 39¢ = ______

\$3.41 + \$0.29 = _______

\$1.26 – 27¢ = _______

Add 376 + 4585.

Add.

	6	3	7	4
+	2	8	1	8

Subtract.

	5	6	3	7
–	2	6	8	9

Solve the problem.

Casey received \$1.03 on Monday and three quarters and four nickels on Tuesday. How much less money did Casey receive on Tuesday? Write your answer two ways using a ¢ and \$.

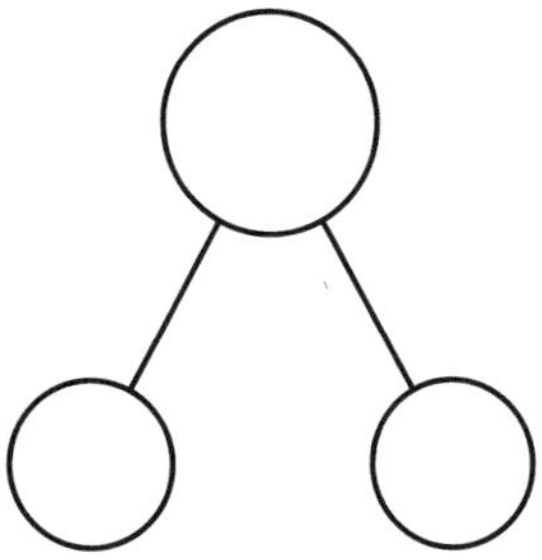

____

Write >, <, or = on the lines.

1 dime _____ 1 nickel

1 dollar _____ 5 quarters

\$0.98 _____ 98¢

42 pennies _____ 9 nickels

3 quarters _____ 7 dimes

Complete the equations.

6 nickels = _____ dimes

_____ nickels = 3 quarters

\$1.62 = \$1 + _____ cents

1 nickel + _____ dimes = 1 quarter

3 dimes + _____ nickels = 2 quarters

Name: ______________________________

Date: ______________________________

Make all measurements in centimeters.

1. Which line is longer, Line A or Line B? _______ How much longer? _______
2. Find the perimeter of the trapezoid. Write the perimeter equations.

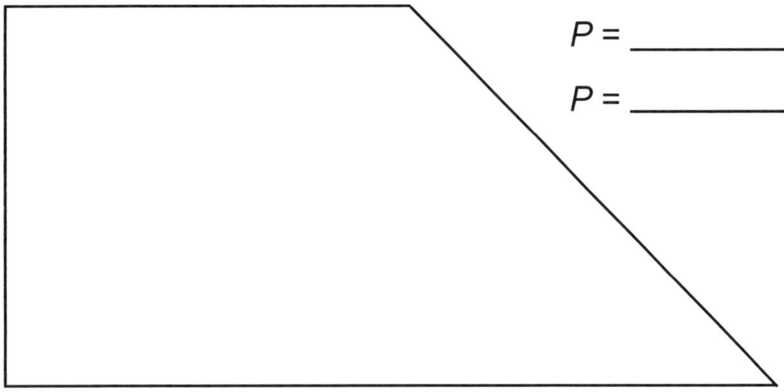

P = ______________________________

P = ____________

Line A

3. Find the perimeter of the regular pentagon. Write the equations two ways using addition and using multiplication.

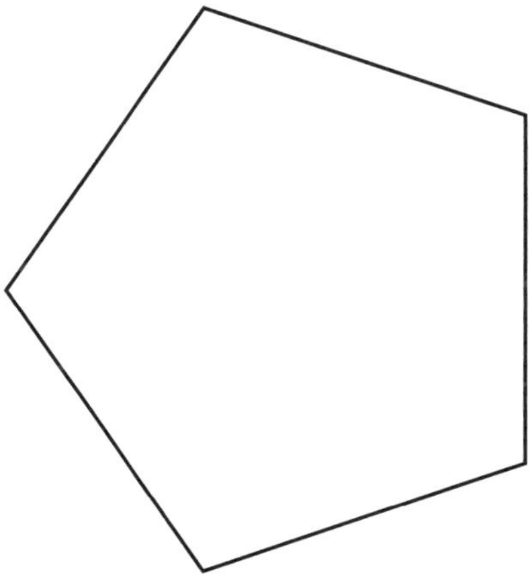

P = ______________________________

P = ____________

P = ____________

Line B

4. Find the perimeter of the regular star. Write the equation and answer.

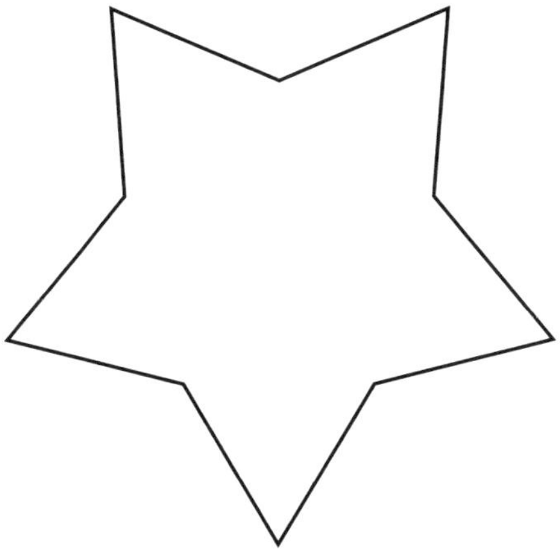

____

____

Name:______________________________ Date: ______________________________

1. Draw a rectangle that measures $7\frac{1}{2}$ cm by 3 cm. Start at the dot. Find the perimeter in centimeters.

•

2. Draw a rectangle that measures $7\frac{1}{2}$ in. by 3 in. Start at the dot. Find the perimeter in inches.

•

Name: ______________________________

Date: ______________________________

Solve the problems.

1. Measure your height and another person's height in inches. Then change the measurements to feet and inches.

______ in. = ______ ft ______ in.

______ in. = ______ ft ______ in.

2. An amusement park has a ride that only children shorter than 1 yard can ride. Can the following children ride:

Ian, 2 ft 6 in. ______ Lily, 3 ft ______ Jordan, 3 ft 3 in. ______

3. A dog is in a backyard that is in the shape of a square. One side of the backyard is 38 feet. How far does the dog walk when it walks around the whole backyard?

____

4. A basketball player is 6 feet 10 inches tall. How tall is that in inches? Explain how you found the number of inches.

____

____

5–10. Use a calculator to change the following to inches.

4 ft 6 in = ________ 2 ft 8 in = ________ 5 ft 3 in = ________

8 ft 7 in = ________ 11 ft 9 in = ________ 10 ft 5 in = ________

11. In most newer houses the ceiling is 8 feet above the floor. How far from the ceiling is the basketball player's head in the fourth problem?

____

Name: ______________________________

Date: ______________________________

Fill in the table. First guess carefully, then measure and find the difference.

Object	Your Guess	Measured	Difference

Name: ______________________________

Date: ______________________________

The rulers in the pictures show inches.

a

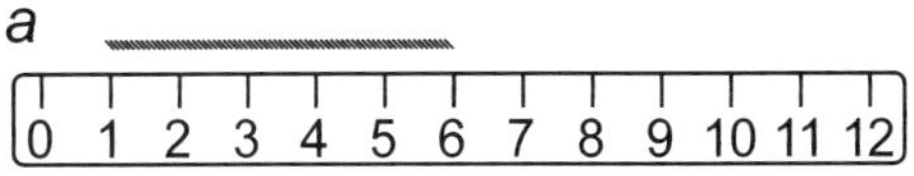

1. How long is string *a*? ________ Explain.

____

____

b

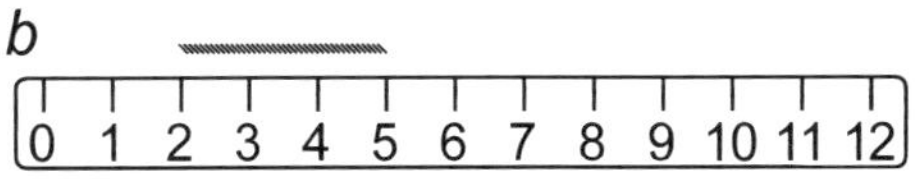

2. How long is string *b*? ________ Explain.

____

____

c

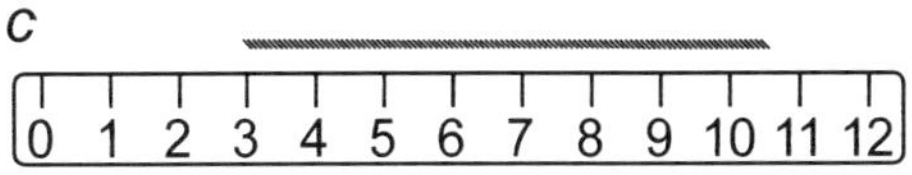

3. How long is string *c*? ________ Explain.

____

____

d

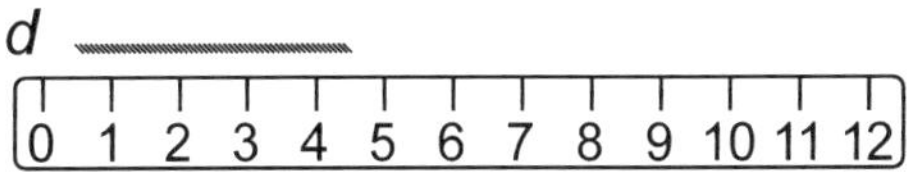

4. How long is string *d*? ________ Explain.

____

____

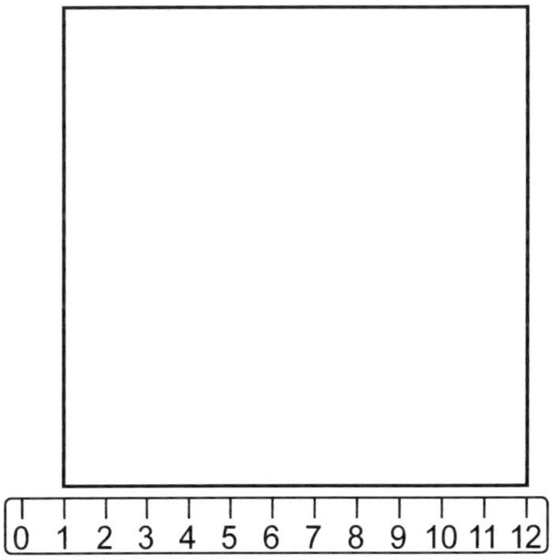

5. Find the perimeter of the square in feet and inches.

____

____

____

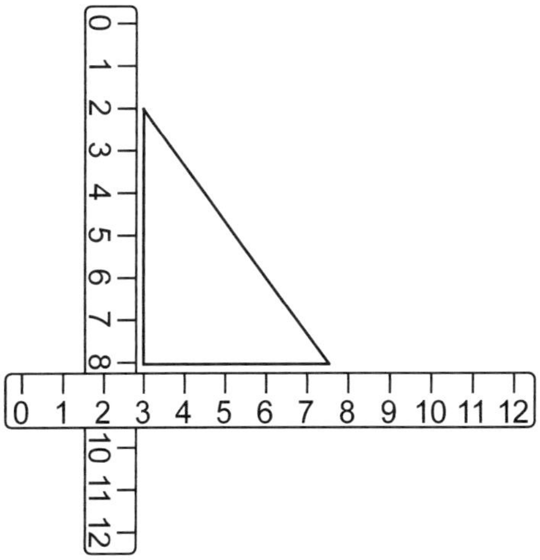

6. The perimeter of the triangle is 18 in. What are the lengths of the three sides?

____

____

____

Name: ______________________________

Date: ______________________________

Draw a 4 cm by 4 cm rectangle. Then draw two more rectangles with the same area. Write the areas and perimeters of all three rectangles, labeling the units.

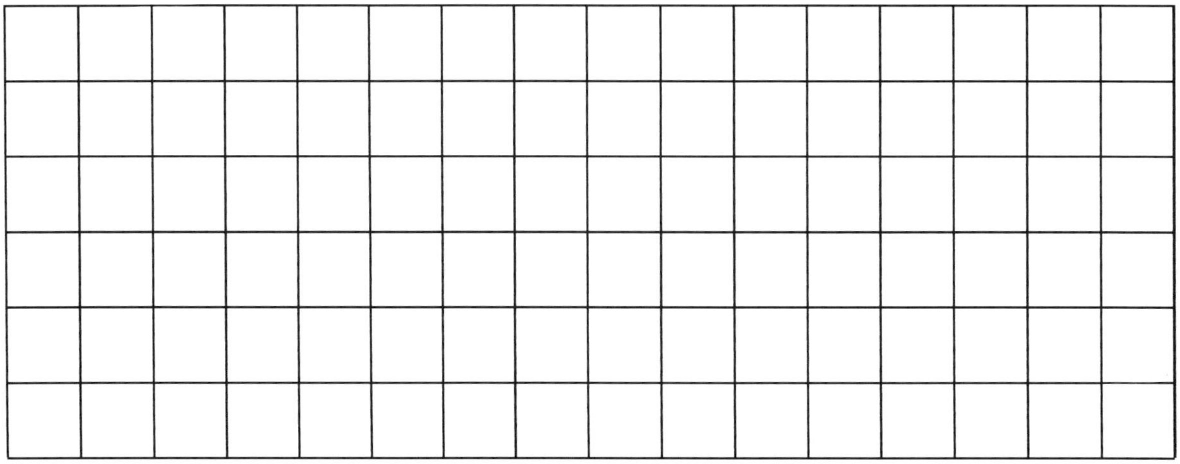

____

____

____

Find the areas of the rectangles. Color or mark the rectangles that have the same areas with the same color or mark. You can write on the rectangles.

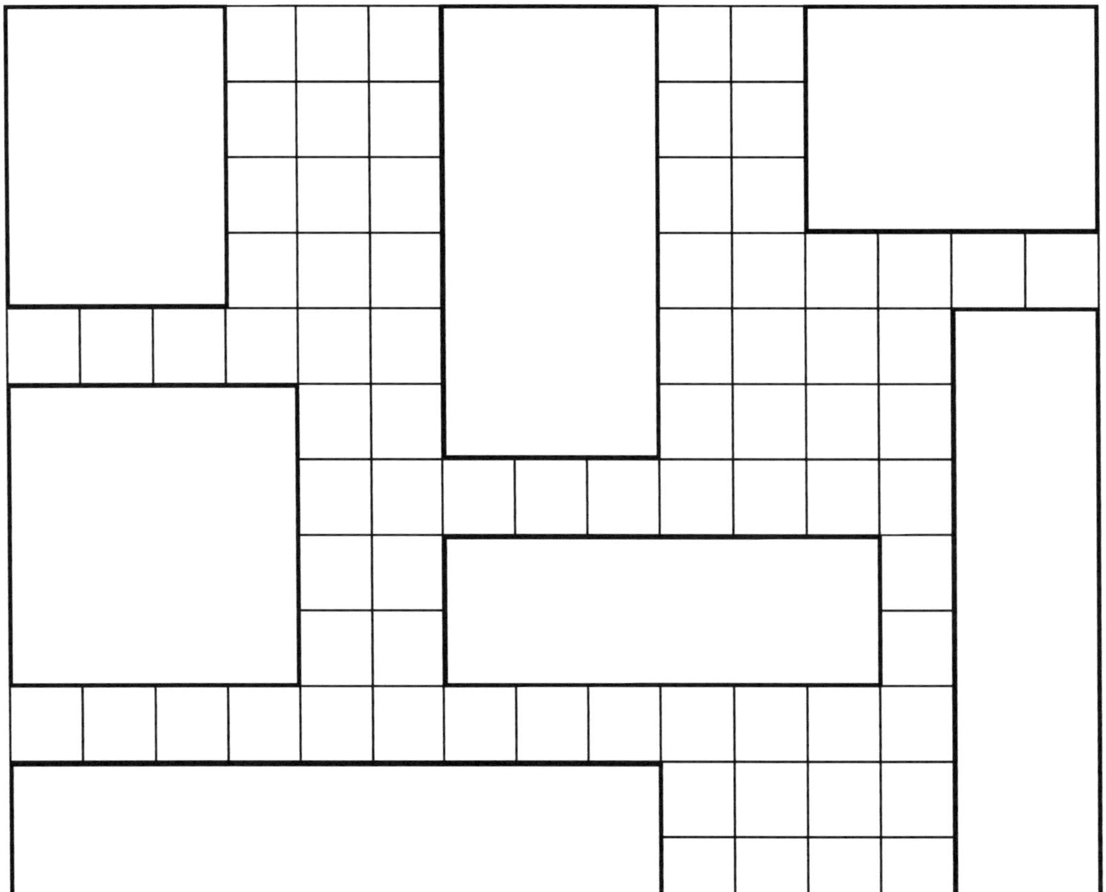

Name: ______________________________

Date: ______________________________

1. Find the areas of the shapes in cm^2. Then draw the six shapes on the grid below in order from least area to greatest area.

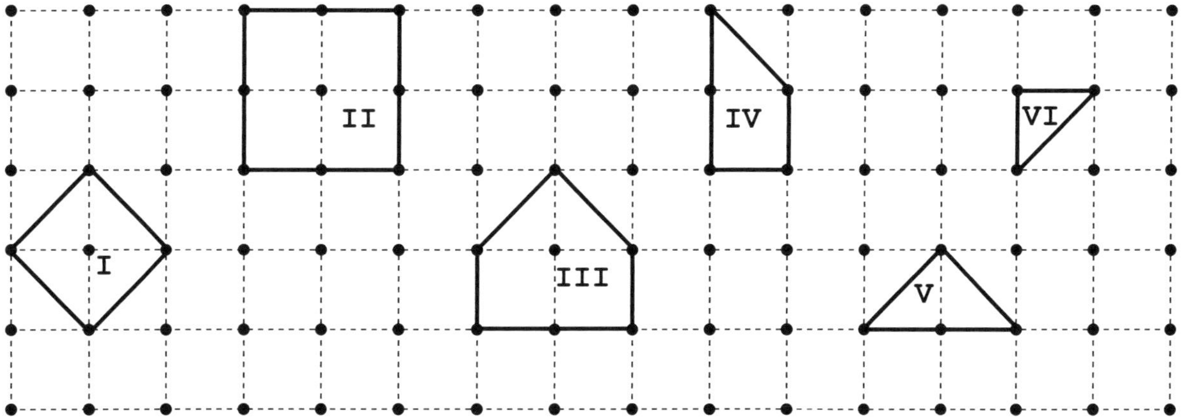

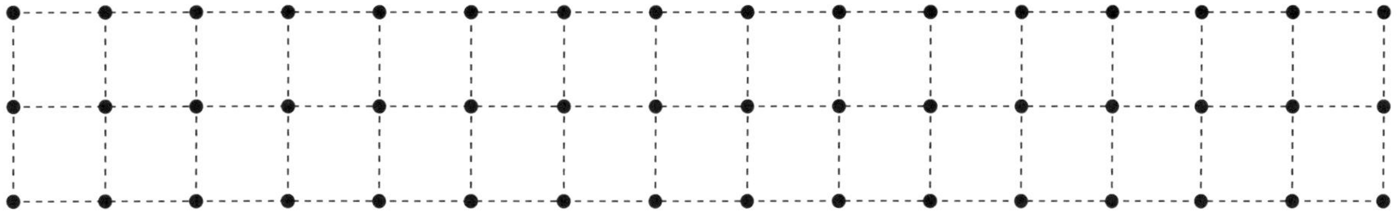

2. Find the areas of the shapes in cm^2 and write them below. Draw your own shape and call it *J.*

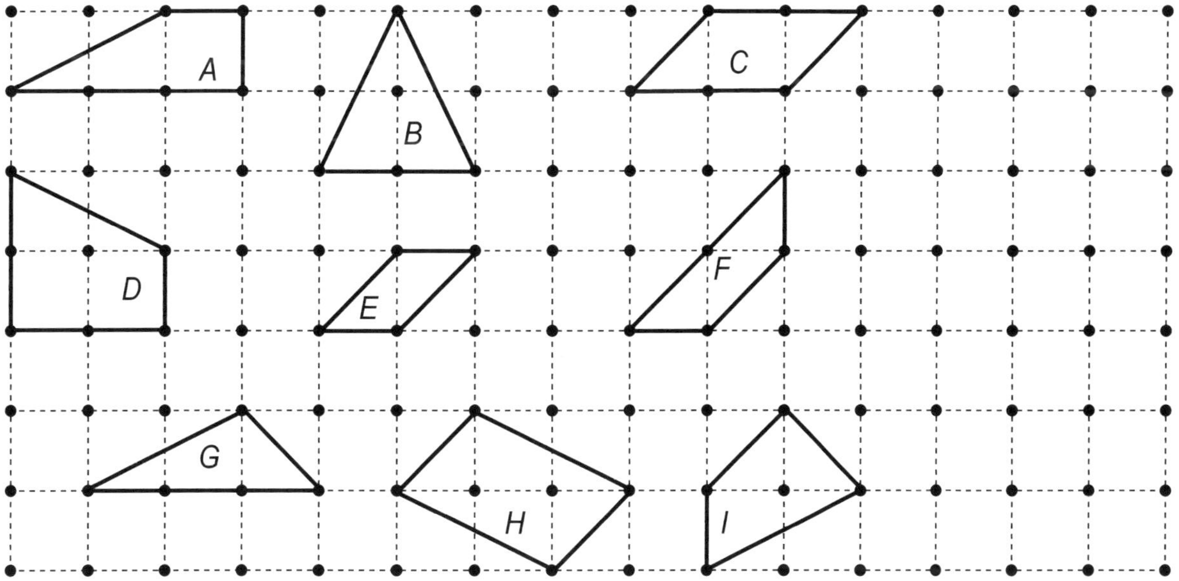

A. ____________________ *B.* ____________________ *C.* ____________________

D. ____________________ *E.* ____________________ *F.* ____________________

G. ____________________ *H.* ____________________ *I.* ____________________

J. ____________________

Name: ______________________________

Date: ______________________________

Write only the answers.

____

____

____

Write the answers.

4 dimes + 17¢ = ______

86 + 17 = ______

5×5 = ______

Solve the problem.

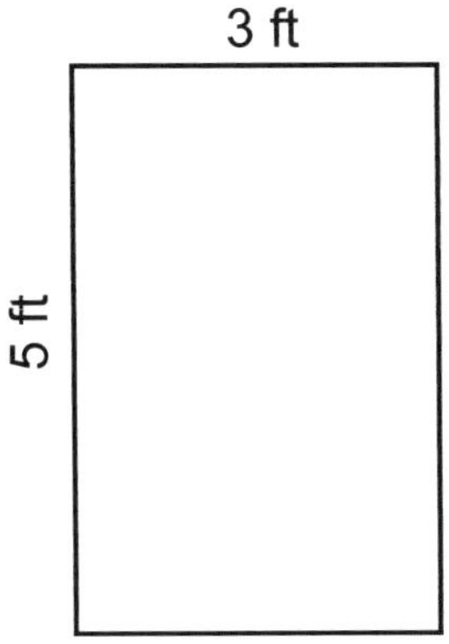

Alex and friends want to make a fence for a pet. They have 4 yards of fencing material. Is it enough to make the pen shown? Explain.

Measuring problem.

Use your centimeter ruler and find the perimeter and area of the rectangle below.

____

____

Write >, <, or = on the lines.

3 ft _____ 1 yard

4 cm _____ 4 in.

24 in. _____ 1 ft 10 in.

1 meter _____ 1 yard

4 in. _____ 10 cm

Complete the equations.

3 ft = _____ in.

_____ in. = 1 ft

1 meter = _____ cm

2 ft + _____ in. = 30 in.

4 ft – _____ in. = 1 yard

Name: ______________________________

Date: ______________________________

Write only the answers.

____

____

____

Write the answers.

5 dimes + 21¢ = ______

74 + 49 = ______

5 × 4 = ______

Solve the problem.

Robin and Morgan made a sandbox for their sister. It has four sides each 1 meter long. What is the distance around the sandbox in centimeters?

Measuring problem.

Use the inches on your ruler and find the perimeter and area of the rectangle below.

____

____

Write >, <, or = on the lines.

1 yard ____ 1 meter

3 in. _____ 3 cm

21 in. _____ 1 ft 9 in.

36 in. _____ 1 yard

2 in. _____ 5 cm

Complete the equations.

3 ft = _____ in.

12 in. = _____ ft

2 meter = _____ cm

10 in. + 1 ft = ___ in.

1 yard – 1 ft = _____ in.

Name: ______________________________

Date: ______________________________

1. Take a handful of colored tiles. Fill in the table for the number of each color. Make the line plot on the line.

____

Which color do you have the most tiles? ___________

What color do you have the fewest tiles? ___________

How many blue tiles do you have? ___________

2. Write the abbreviations for the months of the year. Fill in the table for the number of people having birthdays in that month. Make the line plot on the line below.

Jan	
Feb	
Mar	

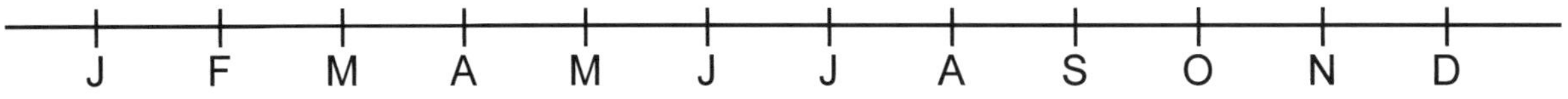

Number of Birthdays in Each Month

What month has the fewest birthdays? ___________

How many birthdays are in June, July, and August? ___________

In what month is your birthday?___________

How many birthdays are in the month of your birthday? ___________

Name: ____________________

Date: ____________________

1. Write the number of facts for each sum in the tables.

Do not use addends greater than 9. Do not include duplicate facts or facts with zeroes.

2. Then make a line plot below with the number of facts.

Sum	Facts
2	1
3	
4	
5	
6	
7	
8	
9	
10	

Sum	Facts
11	4
12	
13	
14	
15	
16	
17	
18	

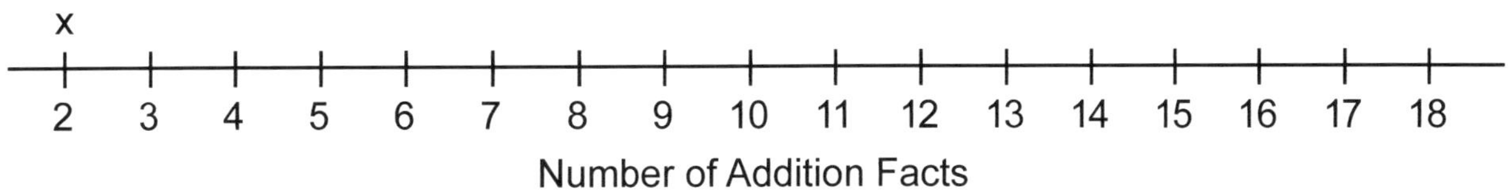

3. Which sum has the greatest number of facts? ______

4. Which sums have four facts? ____________

5. How many facts are on your line plot? ______

6. How many facts have sums greater than ten? ______

7. Are there more facts less than ten or more facts greater than ten? _________

8. Play the Addition War game and record all the sums on the line plot below.

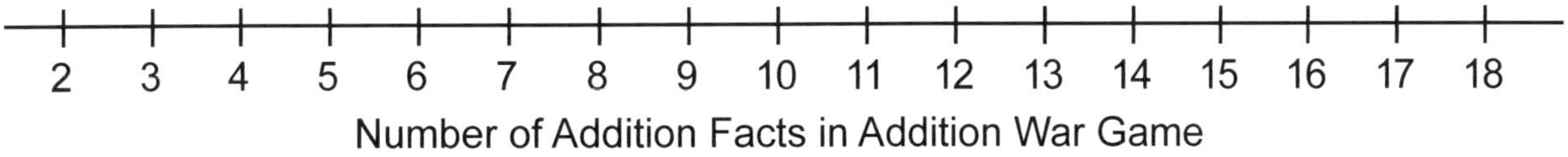

9. How many facts are there on your line plot? ____________________

10. How is this line plot different from the one above? ____________________

____

Name: ____________________

Date: ____________________

1. Use the areas of the shapes from Worksheet 74 to fill in the tables.

2. Then use the data to make a line plot below.

Shape	Area in cm^2
I	
II	
III	
IV	
V	
VI	

Shape	Area in cm^2
A	
B	
C	
D	
E	
F	
G	
H	
I	

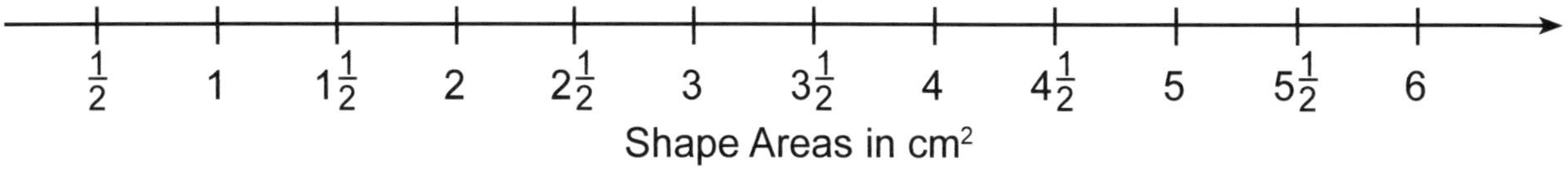

3. Draw at least four shapes on the grid. Include a rectangle, a pentagon, and a hexagon. Fill in the table below.

4. Add the data of these shapes to the line plot above. Underline the x's of your new data.

Shape	Area in cm^2
Rectangle	
Pentagon	
Hexagon	

5. Which shape has the least area? ____________________

6. Which shape has the greatest area? ____________________

7. What area is the most common? ______

8. How many different areas are on your line plot? ______

Name: ____________________ Date: ____________________

First make the shape with tangram pieces. Then draw the shape with your drawing tools on the line below.

1. Draw a square with the two smallest triangles.

2. Draw a parallelogram with the two smallest triangles.

3. Draw a medium triangle with the two smallest triangles.

4. Draw a square with the three smaller triangles.

5. Draw a square with the two largest triangles.

6. Draw a square with the three smaller triangles, the square, and the parallelogram.

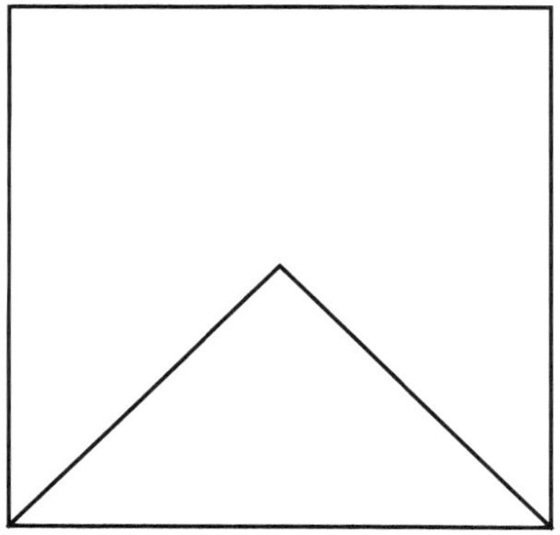

Name:______________________ Date: ______________________

First make each rectangle with your tangram pieces. Then complete the rectangles with your drawing tools.

1. Draw a rectangle by adding two small triangles.

2. Draw a rectangle by adding a small and a medium triangle.

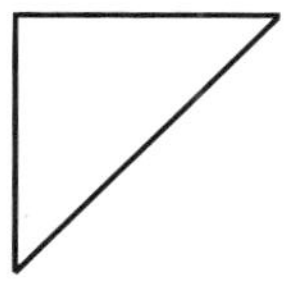

3. Draw a rectangle by adding two small triangles and a medium triangle.

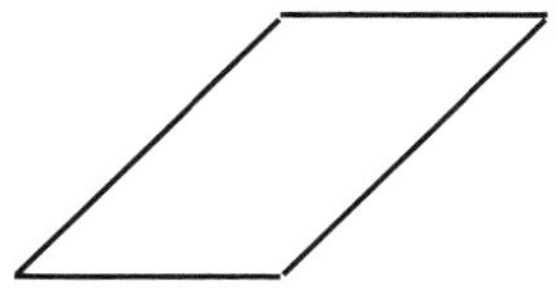

4. Draw a rectangle by adding the three smaller triangles.

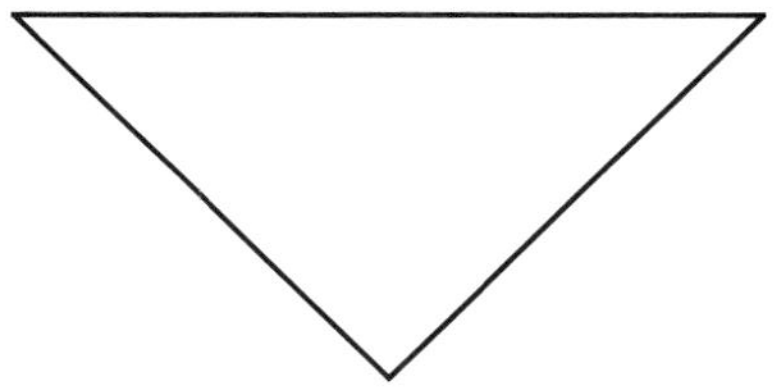

5. Draw a rectangle by adding four triangles.

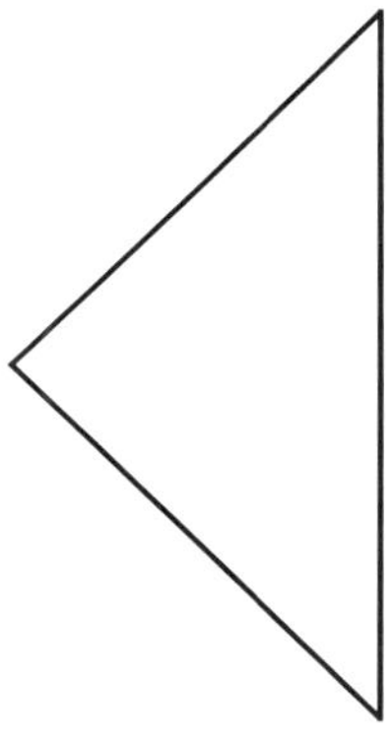

6. Draw a rectangle by adding three triangles. (The rectangle sides will not be parallel to the paper.)

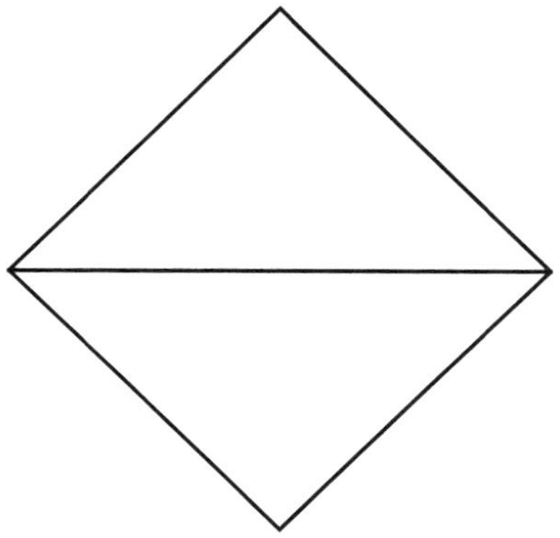

7. Draw a rectangle by adding a large triangle, a small triangle, and a parallelogram.

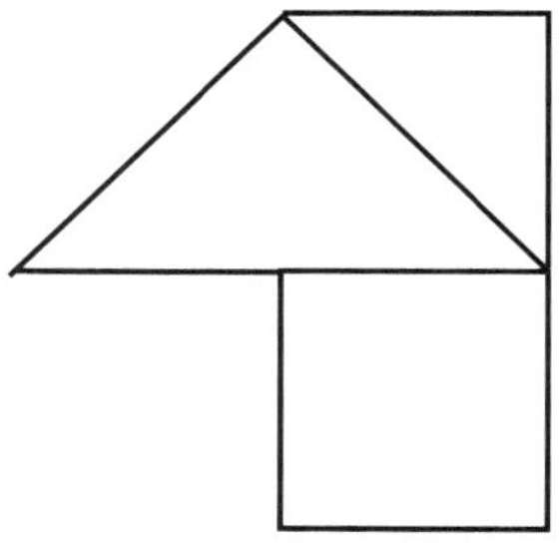

Name: ______________________________

Date: ______________________________

Find tangram pieces that fit in the large shaded trapezoid. Then draw the solutions below with your drawing tools.

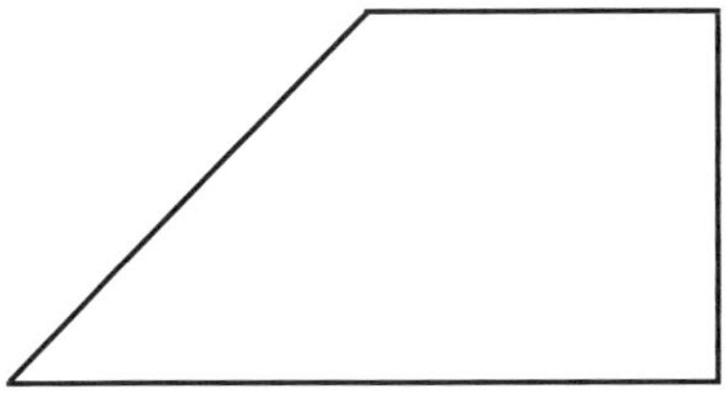

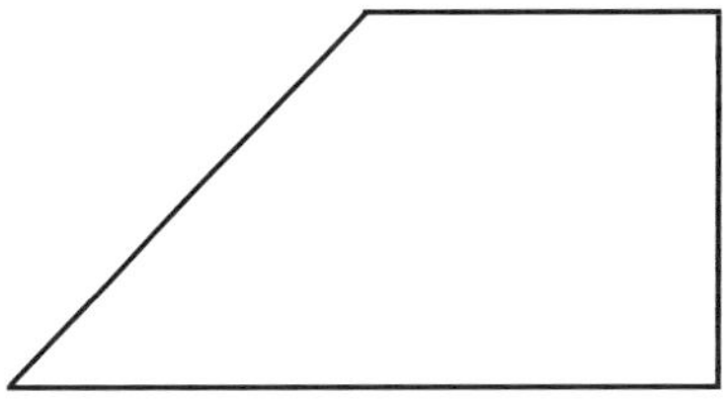

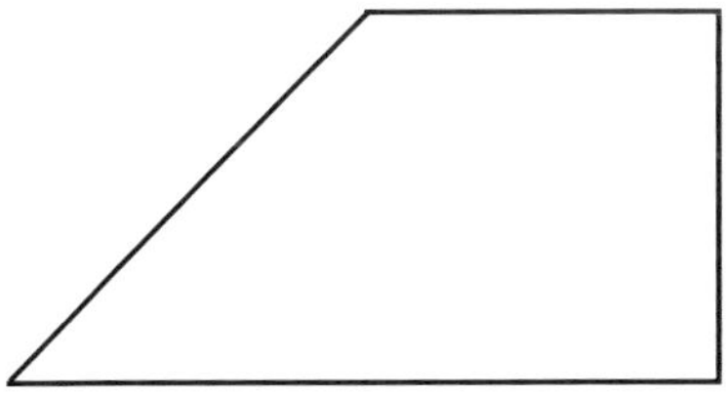

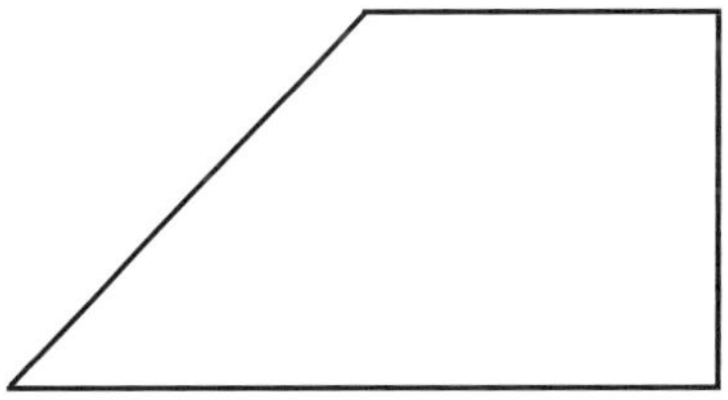

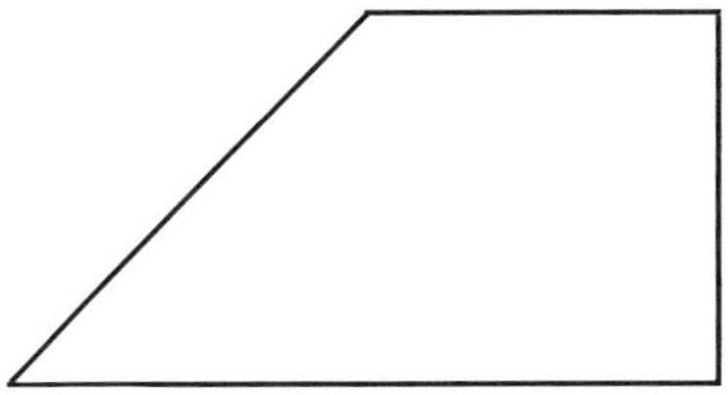

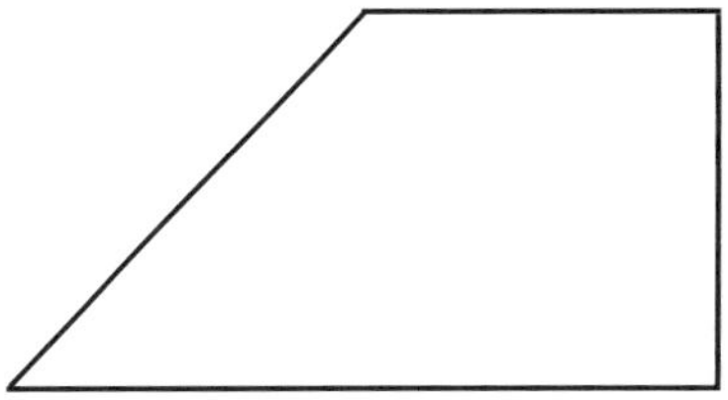

Name:____________________________ Date: ____________________________

Use your drawing tools to make a reflection of the tangram pieces. The dashed line is the line of symmetry.
If you would like, first make the reflection with tangram pieces. Check your work with a reflector.

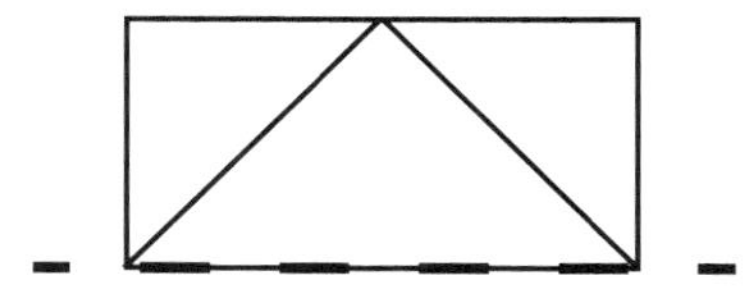
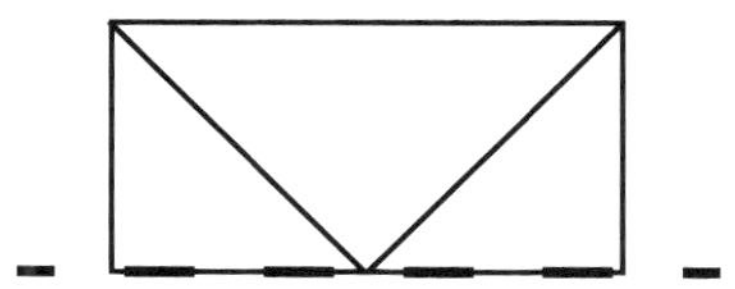
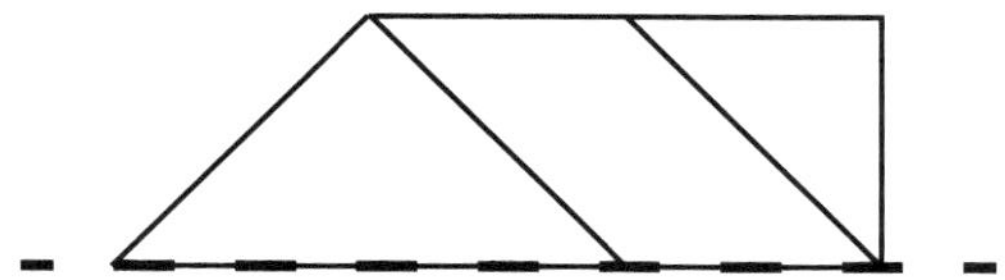

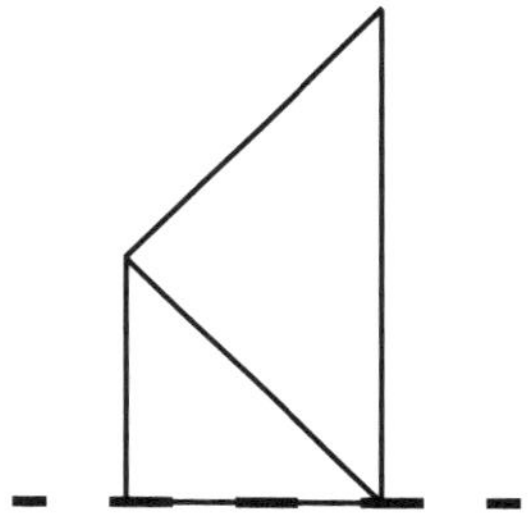
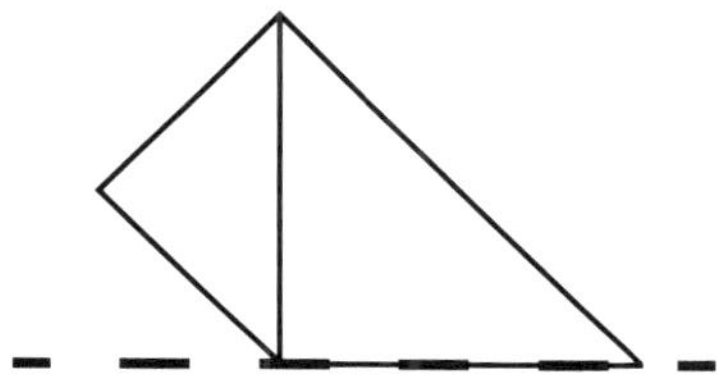
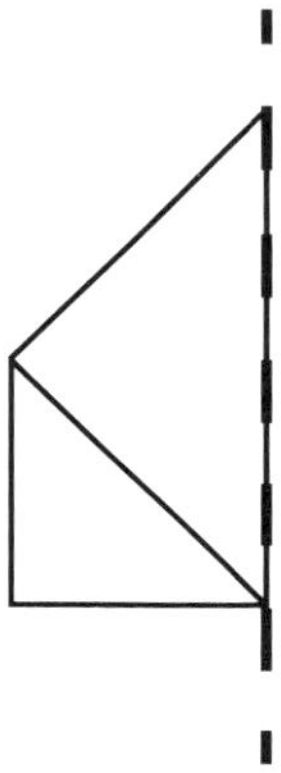
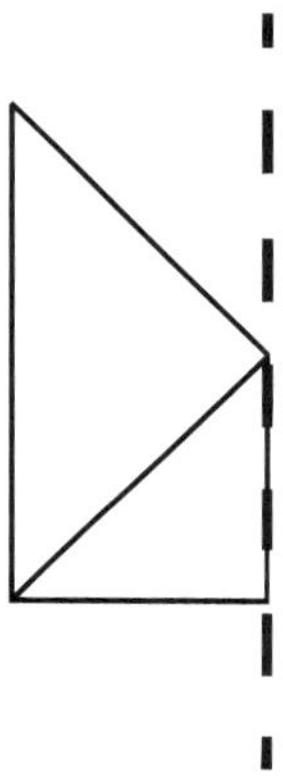

What do you notice about the figures in the second row? ____________________________

____

Name: ______________________________

Date: ______________________________

Use your abacus to find the missing factor.

6 × 3 = 18	7 × 4 = 28	12 = 6 × ____
5 × 6 = 30	5 × 3 = 15	20 = 4 × ____
8 × 7 = 56	5 × 10 = 50	72 = 9 × ____
9 × ____ = 45	8 × ____ = 48	72 = 8 × ____
5 × 9 = 45	6 × ____ = 48	32 = 8 × ____
6 × ____ = 54	8 × ____ = 64	25 = 5 × ____

Use your abacus to find the missing factor.

____ × 4 = 32	____ × 5 = 35	32 = ____ × 8
____ × 3 = 18	____ × 6 = 42	12 = ____ × 2
____ × 10 = 40	____ × 7 = 42	36 = ____ × 6
____ × 7 = 35	____ × 4 = 20	72 = ____ × 8
____ × 9 = 81	____ × 6 = 18	72 = ____ × 9

Name: ______________________________

Date: ______________________________

Write and solve the missing factor equations. Complete the part-whole circle sets.

1. Mickey has 24 ounces of lemonade and is pouring 6 ounces into each glass. How many glasses of lemonade can Mickey fill?

____

2. Dana has 24 ounces of chocolate milk and is pouring the same amount into 3 glasses. How much can Dana pour into each glass?

____

Divide. Use your abacus when you need it.

$\frac{8}{4} =$ ____	$\frac{12}{4} =$ ____	$\frac{28}{4} =$ ____
$\frac{10}{2} =$ ____	$\frac{15}{5} =$ ____	$\frac{25}{5} =$ ____
$\frac{20}{5} =$ ____	$\frac{40}{8} =$ ____	$\frac{30}{5} =$ ____
$\frac{16}{4} =$ ____	$\frac{20}{4} =$ ____	$\frac{48}{8} =$ ____

Name: Jaxon

Date: 11/19/24

Circle $\frac{1}{2}$ of each figure.

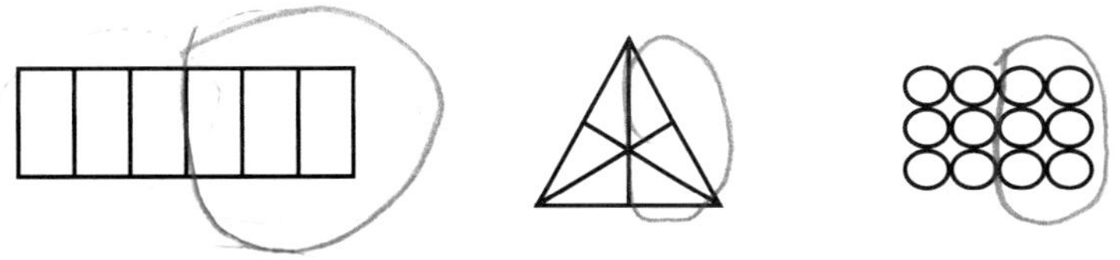

Circle $\frac{1}{3}$ of each figure.

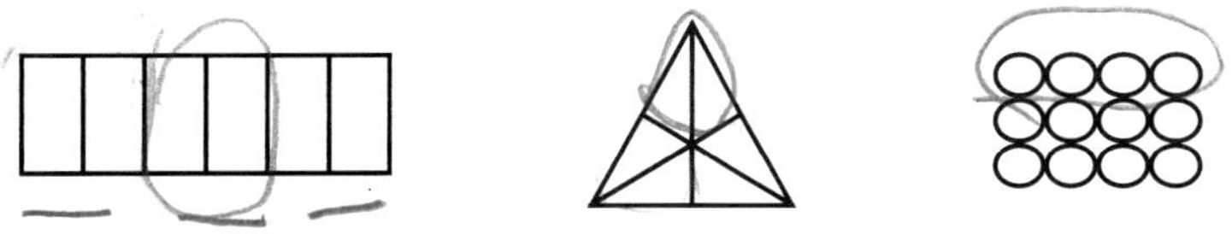

Circle $\frac{1}{4}$ of each figure.

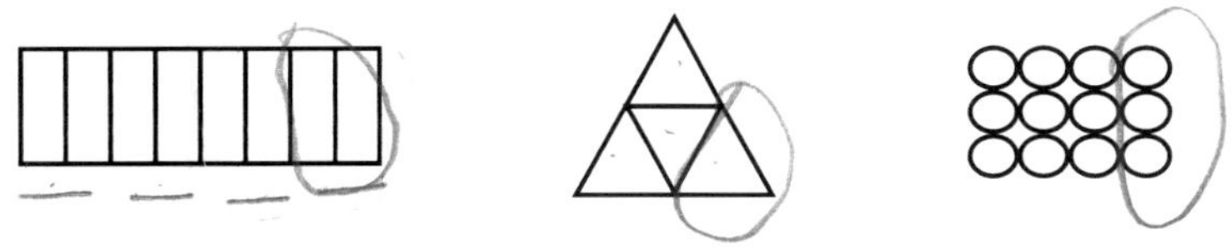

Draw lines to match unit fractions to the figure that shows the fraction with shaded area.

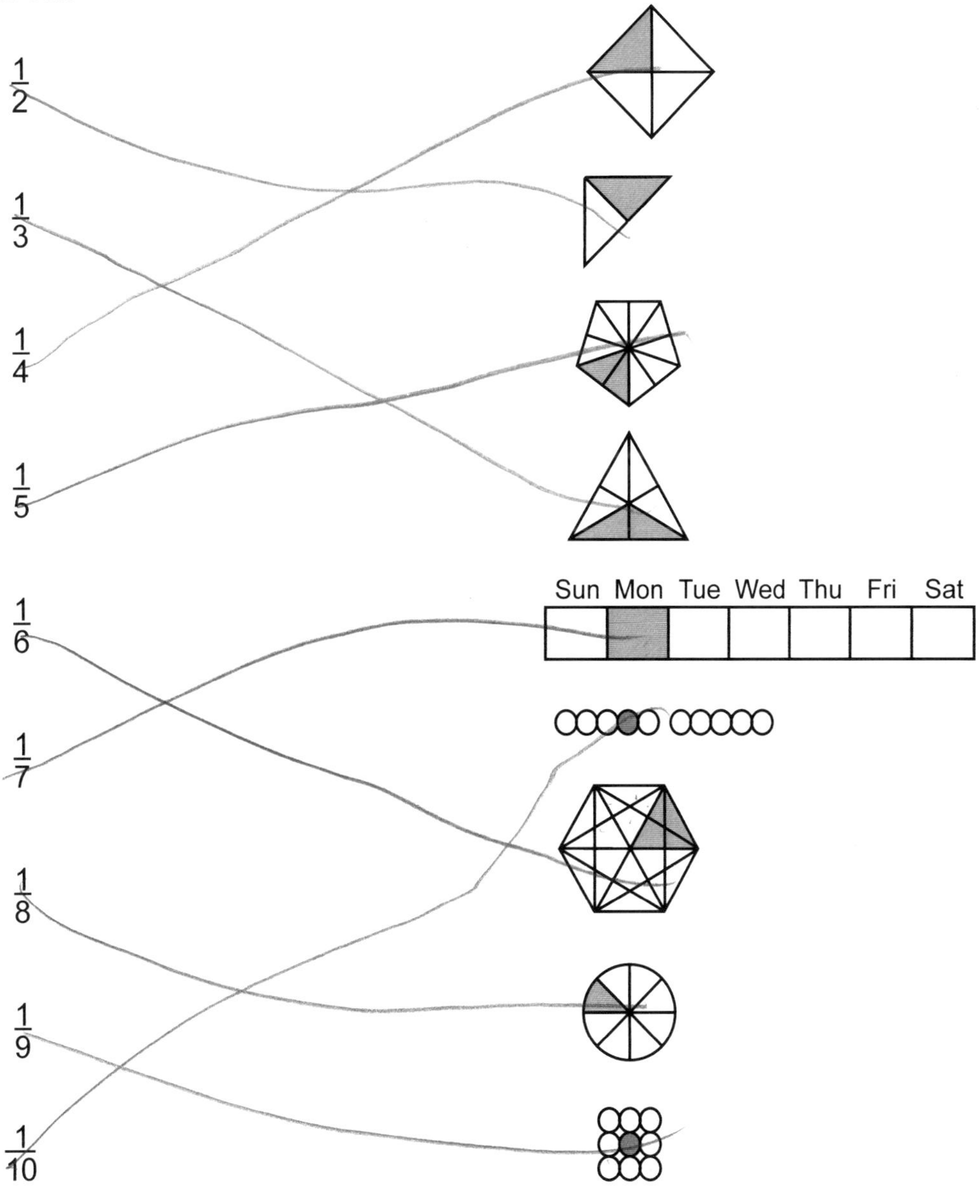

Name: ______________________________

Date: ______________________________

Write the fractions that are circled in each row.

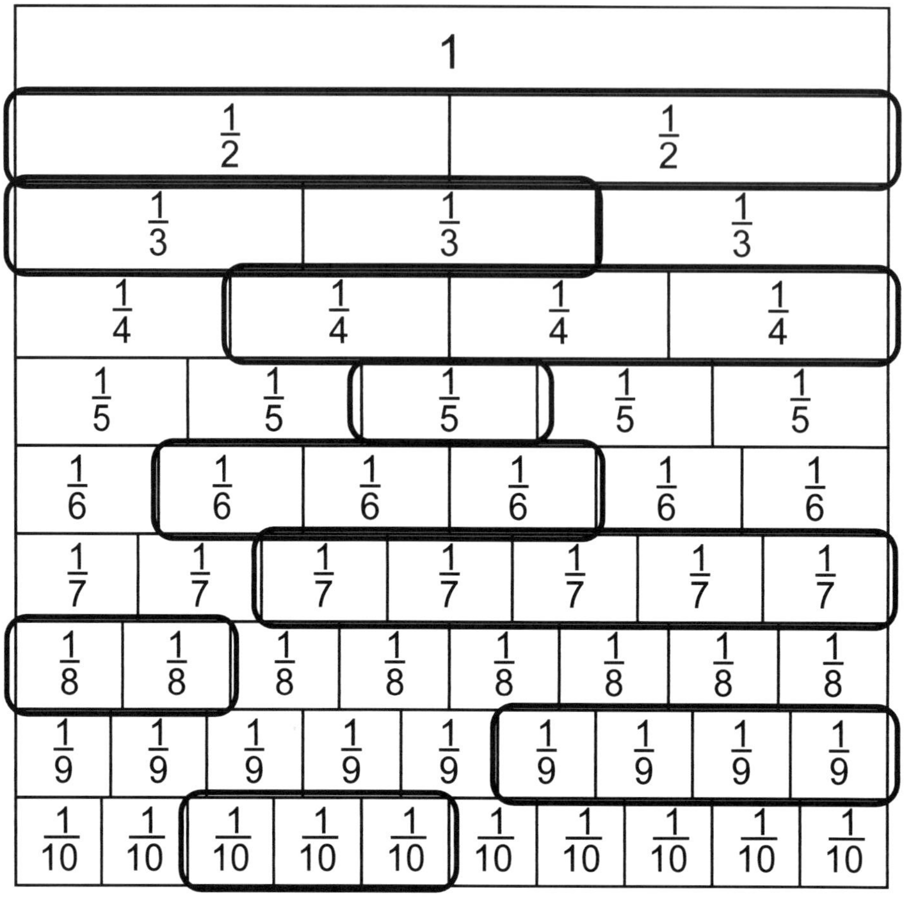

______

______

______

______

______

______

______

______

______

Match the fractions that will be equal to one.

$\frac{2}{4}$	$\frac{4}{6}$
$\frac{2}{10}$	$\frac{8}{10}$
$\frac{2}{6}$	$\frac{2}{4}$
$\frac{4}{8}$	$\frac{5}{10}$
$\frac{6}{9}$	$\frac{4}{8}$
$\frac{2}{8}$	$\frac{6}{8}$
$\frac{5}{10}$	$\frac{3}{9}$

Name: ______________________________

Date: ______________________________

1. There are 3 granola bars to be split equally among 4 friends. How many fourths does each friend get? Explain your work.

Cut out the three rectangles at the right to represent the bars.

2. Four children have three small pizzas to divide among them. Explain with pictures and words how much each child gets.

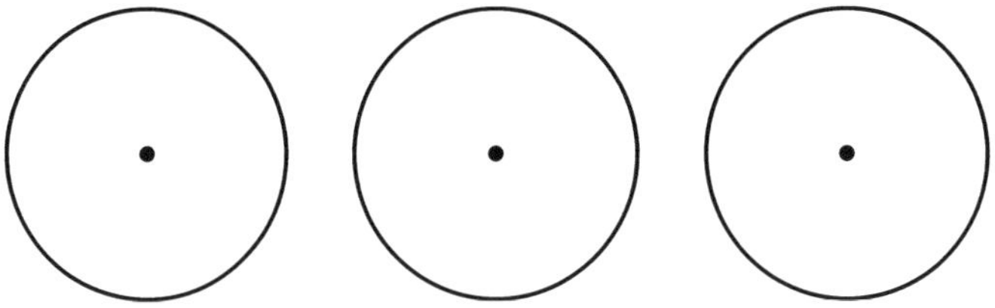

Name: ______________________________

Date: ______________________________

Make the following shapes with the Geometry Panels.

Triangular prism

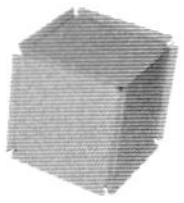

Rectangular prism (cube)

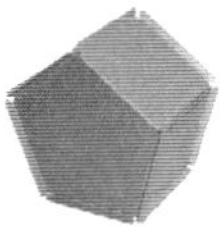

Pentagonal prism

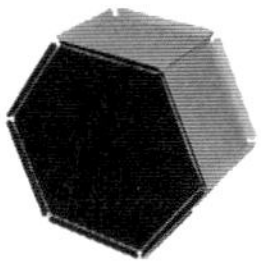

Hexagonal prism

Triangular pyramid (tetrahedron)

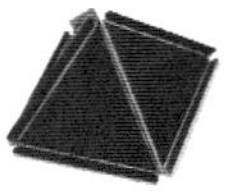

Square pyramid

Pentagonal pyramid

Fill in the table below.

Shape	Faces	Edges	Vertices
triangular prism			
rectangular prism			
pentagonal prism			
hexagonal prism			
triangular pyramid			
rectangular pyramid			
pentagonal pyramid			

Write the names of the solids below. Use the names listed in the table above.

 ______________________________

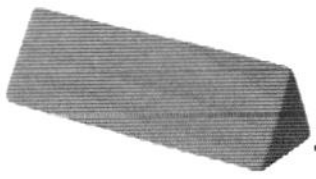 ______________________________

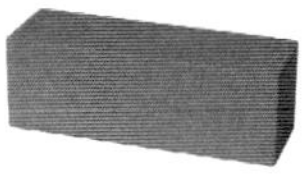 ______________________________

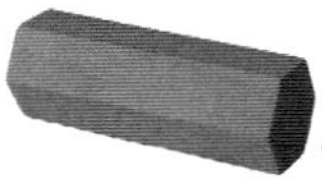 ______________________________

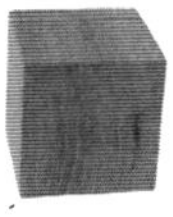 ______________________________

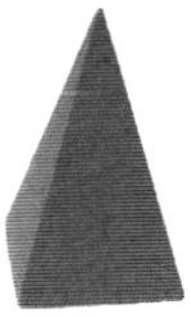 ______________________________

 ______________________________

Name: ______________________________

Date: ______________________________

1. Compared to the centimeter cube, how many times larger do you think the wooden cube is? ________

2. Compared to the wooden cube, how many times larger do you think the yellow cube is? ________

3. Compared to the centimeter cube, how many times larger do you think the yellow cube is? ________

4. Fill in the chart.

Cube	Edge Length	Face Area	Cube Volume
centimeter cube			
wooden cube			
yellow cube			

5. Compared to the centimeter cube, how many times larger is the wooden cube? ________

6. Compared to the centimeter cube, how many times larger is the yellow cube? ________

7. Look at the wooden cube and the yellow cube. How many wooden cubes would fit in the yellow cube? ________

8. Compared to the wooden cube, how many times larger is the yellow cube? ________

9. The yellow cube holds a liter. What fraction of a liter is the wooden cube? ________

Name: ______________________________

Date: ______________________________

Draw a 9 by 2 rectangle on the grid. Then draw another rectangle that has the same area. Write the areas and perimeters of both rectangles, labeling the units.

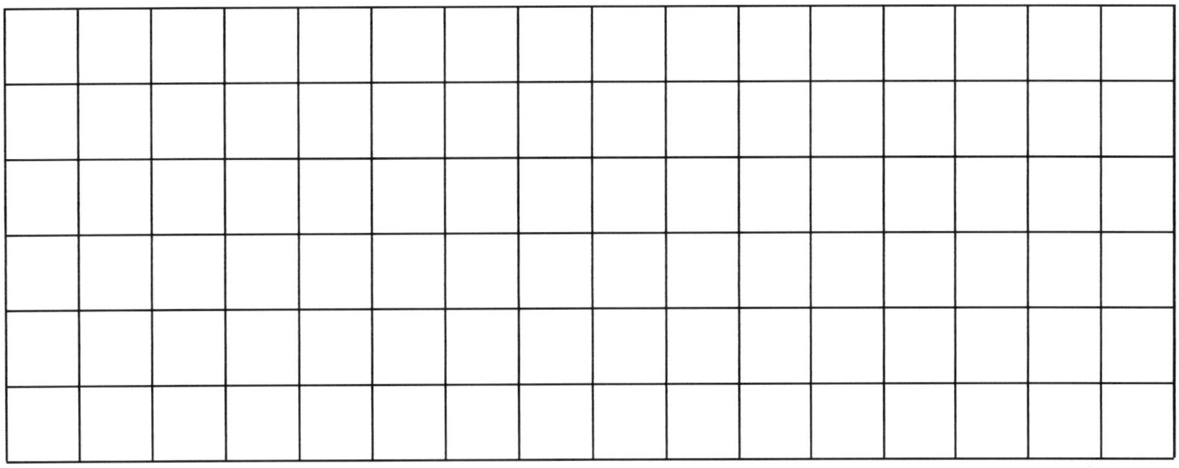

____

____

Circle $\frac{1}{2}$.

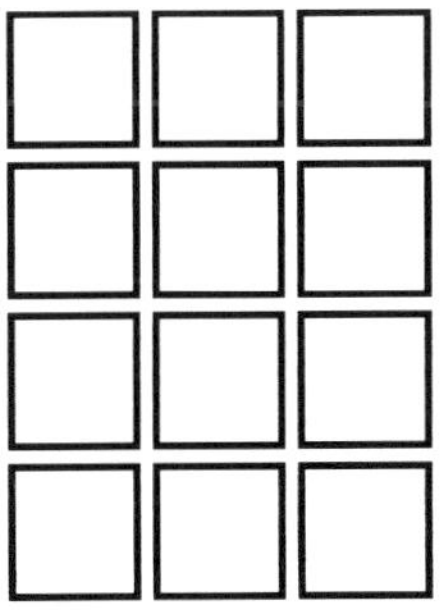

Circle $\frac{1}{3}$.

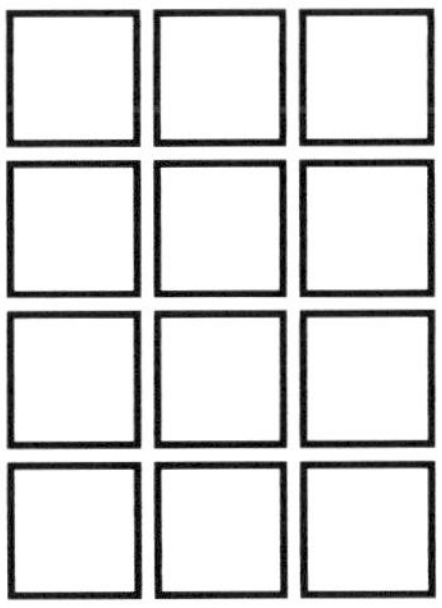

Circle $\frac{1}{4}$.

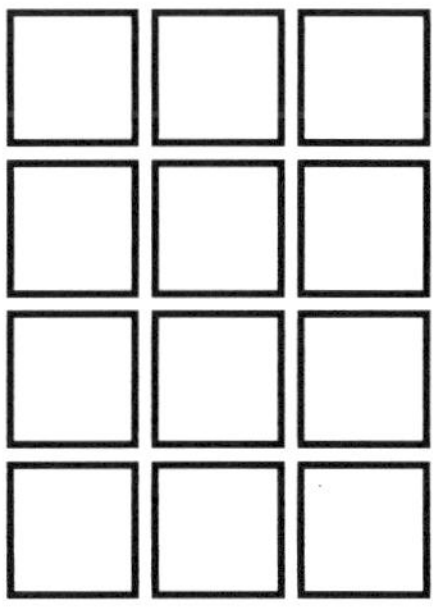

Name: ______________________

Date: ______________________

Estimate what Lines A and B on the sides of your worksheet measure and write your guess in the chart. Then use your ruler to measure the lines and find the difference.

	Your Guess	Measured	Difference
Line A in centimeters			
Line A in inches			
Line B in centimeters			
Line B in inches			

Draw lines to match the digital clocks to the analog clocks.

Line A

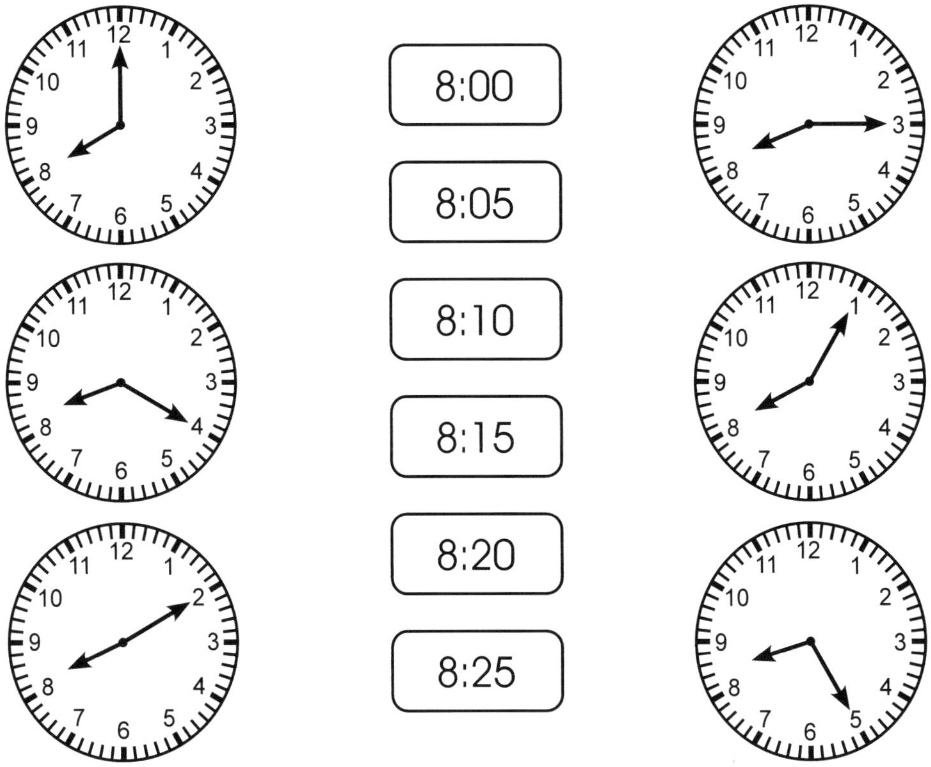

Line B

Read the following and then fill in the line plot below.

Kat has 3 radishes, 3 potatoes, 2 carrots, and 1 squash for sale. Kim has 1 potato, 2 carrots, 2 radishes, and 5 squash for sale. What is the total amount of produce that is for sale?

Potatoes Carrots Squash Radishes

____

Name: ____________________

Date: ____________________

Write the expanded form of the base-10 picture cards. Then write the number in standard form.

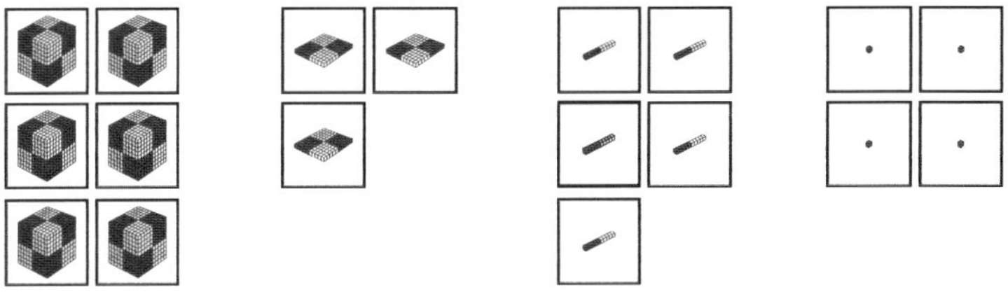

_ 0 0 0 + _ 0 0 + _ 0 + _   _ _ _ _

_ 0 0 0 + _ 0 0 + _ 0 + _   _ _ _ _

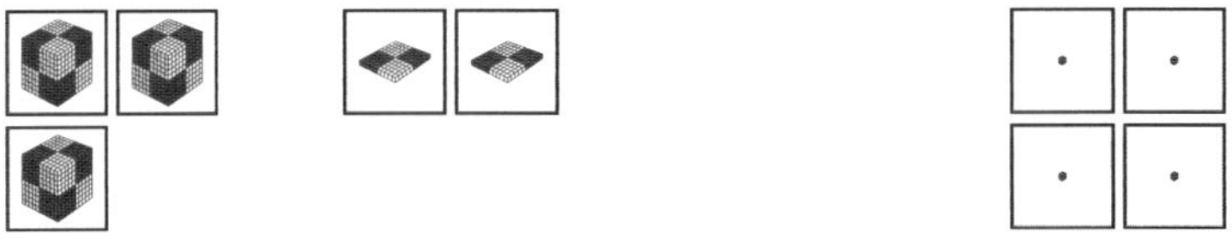

_ 0 0 0 + _ 0 0 + _ 0 + _   _ _ _ _

Use the <, >, or = signs to finish the equations.

321 + 100 ____ 411 + 10

98 + 63 ____ 100 + 85

485 + 10 + 1 ____ 100 + 336

460 + 10 ____ 9 + 461

Solve.

	4	3	5	6
+	2	3	4	5

	9	5	6	5
+	1	2	4	5

	5	8	7	6
–	3	9	6	7

	7	2	6	5
–	3	6	8	9

15 + 12 + 45 + 18 = ______

64 + 34 + 16 + 35 = ______

41 + 56 + 80 + 23 = ______

32 + 50 + 64 + 36 = ______

Math Journal

Math Journal

Math Journal

MATH JOURNAL

MATH JOURNAL

Math Journal

Math Journal

Math Journal

Math Journal

MATH JOURNAL

Math Journal

Math Journal

Math Journal

Math Journal

Math Journal

Math Journal

MATH JOURNAL

Math Journal

Math Journal

Math Journal

Math Journal

Math Journal

Math Journal

MATH JOURNAL

Math Journal

Math Journal

Math Journal

Math Journal

Math Journal

Math Journal

MATH JOURNAL

WARM-UPS

Name: Jaxon

Date: 9/4/24

Subtract. Use your abacus to help.

1.

	7	9	4	9
–	2	7	4	5
	5	2	0	4

2.

	7	9	8	2
–	1	9	2	9
	6	0	5	3

3.

	6	6	5	1
–	5	9	2	8
	0	7	2	3

4.

	5	0	5	7
–	3	6	1	7
	1	4	4	0

5.

	8	1	5	7
–	6	5	6	1
	1	5	9	6

6.

	9	6	1	4
–	5	1	4	6
	4	4	6	8

7.

	3	3	3	9
–	1	3	7	7
	1	9	6	2

8.

	6	5	1	2
–	1	1	1	3
	5	4	0	9

9.

	1	3	1	4
–	1	2	8	2
	0	0	3	2

10.

	3	9	2	4
–	3	5	8	5
		3	3	9

11.

	8	3	2	3
–	4	6	5	7
				6

12.

	7	0	5	0
–	2	4	6	8

Name: ____________________

Date: ____________________

Solve. Use your abacus only if you need it.

1.

	8	7	3	8
+	1	6	3	5

2.

	6	8	6	1
–	2	8	3	9

3.

	7	7	4	3
–	5	8	1	7

4.

	6	0	5	7
+	4	5	3	7

5.

	7	2	5	1
–	2	4	3	2

6.

	9	6	2	3
+	4	1	6	5

7.

	4	4	4	8
+	1	3	6	6

8.

	4	6	2	2
–	3	3	6	5

9.

	1	6	2	3
+	1	4	5	3

10.

	3	8	1	4
–	3	2	7	7

11.

	7	2	1	4
–	3	5	4	8

12.

	7	0	5	0
–	1	3	7	6

Name: ______________________

Date: ______________________

1.

3	15	13	18
7	5	23	9
+ 81	15	33	12
	+ 65	+ 43	+ 31

2.

	1	0	0	0
	–	9	9	9

1	0	0	0	0
–	9	9	9	9

3.

399 + 19 = ______ 372 + 108 = ______

543 + 21 = ______ 293 + 40 = ______

146 + 18 = ______ 731 + 60 = ______

4.

87 + ______ = 100 437 + ______ = 458

243 + ______ = 253 372 + ______ = 400

689 + ______ = 799 731 + ______ = 830

5.

How many days are in a week? = ______
How many months are in a year? = ______
How many hours are in a day? = ______
How many minutes are in an hour? = ______
How many minutes are in a half hour? = ______
How many minutes are in a quarter hour? = ______